High-Tech Society

By the same author

The Labour Party and the Working Class (1976)
The Microelectronics Revolution (ed.) (1980)
The Information Technology Revolution (ed.) (1985)
The Materials Revolution (ed.) (1988)

High-Tech Society

The Story of the Information Technology Revolution

TOM FORESTER

The MIT Press
Cambridge, Massachusetts

In memory of Mum

First MIT Press paperback edition, 1988
First MIT Press edition, 1987

Published in Great Britain by Basil Blackwell Ltd

Printed and bound in Great Britain

Library of Congress Cataloging-in-Publication Data

Forester, Tom.
High-Tech society.

Bibliography: p.
Includes index.
1. Computers and civilization. 2. Computers.
I. Title
QA76.9.C66F66 1987 303.4′834 86-27499
ISBN 0-262-06107-4 (hb)
ISBN 0-262-56044-5 (pb)

Contents

Preface and Acknowledgments

Ever since I first got interested in the computer revolution and its implications for society, I've been waiting for the "definitive" book on the subject to appear. It never did – so finally I decided to try and write it myself.

I doubt whether I have succeeded; indeed, I wonder if anyone will in this fast-changing field. But I think I have produced a useful, fairly concise summary of the information technology story so far, which picks out the essential points, happenings and trends from the wealth of material available and highlights the most important studies, books and reports which have been produced.

Three key objectives were to make *High-Tech Society* readable, comprehensive and balanced in its approach. I have tried to make the subject interesting and accessible to all by tackling it chronologically in the form of various sub-plots; I have tried to make it informative – every acronym or computer term which crops up is defined and explained – without allowing it to become a dry, didactic textbook; and I have tried to steer a middle course between the wide-eyed enthusiasm of the high-tech Pollyannas and the doom-laden pessimism of the computer Cassandras – at the same time eschewing the exaggerated hype and "revelation by re-labeling" characteristic of some writers in the field.

This book is a popular overview and as such draws on secondary sources. I have used material already published in newspapers, magazines and journals such as *Business Week*, *Fortune* (including the International Data Corporation's White Papers to Management), *High Technology*, *Technology Review* and the *Financial Times* and *The Sunday Times* of London. In addition, I would like to acknowledge debts to *Science*, *Scientific American*, *New Scientist*, *FUTURES*, *Omega*, the *Harvard Business Review* and the *Sloan Management Review*. I'm particularly grateful to Michael Marien's *Future Survey*, published by the World Future Society, which is an invaluable service.

Finally, I would like to acknowledge an enormous debt to my wife Delia and children Frances, Jack and Heather, who put up with so much while the book was being written.

Tom Forester
Brighton, December 1986

1 Introduction

The High-Tech Revolution – High-Tech Fever

The High-Tech Revolution

The nations of the world are caught up in a revolution: a technological revolution, which is bringing about dramatic changes in the way we live and work – and maybe even think. This book tells the story of that revolution and assesses its impact on business and society so far. Three broad economic, technological and political trends provide the main driving forces behind the high-tech revolution. Their coming together in the 1980s is resulting in an explosion of technological and social innovation on a scale not seen for many decades, perhaps not since the Industrial Revolution two hundred years ago.

In the first place, the cost of computing power and memory continues to decline rapidly, thanks to developments in microelectronics, once described as "the most remarkable technology ever to confront mankind" (Sir Ieuan Maddock, in Tom Forester (ed.), *The Microelectronics Revolution*, Basil Blackwell, Oxford and MIT Press, Cambridge, Mass., 1980). The microchip has put cheap computing power on the desks of millions. Computers are proliferating as never before: there are now at least 100 million in the world, over half of them born this year and last. Computers have entered into society's bloodstream – and they are becoming ubiquitous because they are cheap.

Second, the digitization of information through the common language of the binary code is bringing about the convergence of voice, image and data – and of the telecommunications, electronics and computing industries based upon them. The binary code is a universal language, a digital language that bridges national frontiers. As Dr An Wang, founder of Wang Laboratories, put it, "The digitization of information in all its forms will probably be known as the most fascinating development of the 20th century."

Third, the worldwide wave of deregulation and the privatization of public monopolies by governments, especially in the field of telecommunications, has sparked off an explosion of entrepreneurial and corporate activity designed to take advantage of the new business environment. New companies enter and leave product areas as never before. New products emerge as never before. Established companies don't know what business they are in, as the boundaries between traditional sectors have become increasingly blurred. As a result, we are witnessing the emergence of an international, integrated information-processing industry based upon digital technology.

As an illustration of the declining cost of computing power, a 1977 state-of-the-art personal computer processed 100,000 machine instructions per second and came with 64 kilobytes of main memory and 160 kilobytes of disk storage. A 1987 machine not only costs much less, but also has 20 times the processing power, 20 times the main memory and 500 times the disk storage capacity. The price of storing a single digital unit of data in a memory chip fell from one-tenth of a cent in 1976 to one-thousandth of a cent in 1986, and it will keep dropping by about 35 percent per annum.

With the new "superchips" (see figure 1.1), we are entering the sub-micron era, with over 1 million transistors placed on a single chip. By the mid-1990s semiconductor designers will be putting over 4 million transistors on a chip, and before the end of the century gigabit – 1 *billion* components on a chip – integration may be possible. The new superchips will probably replace mainframes in time and will soon contain more power than today's supercomputers.

	1980	1985	1987	1990	1995
Circuit size	4 Microns	2 Microns	1 Micron	0.5 Micron	0.25 Micron
Memory capacity	64K	256K	1024K	4096K	16384K
Power range	Desk top computer	Minicomputer	Mainframe computer	Supercomputer	Ultracomputer
Applications	Digital watches, video games, personal computers	Lap computers, Engineering work stations, programmable appliances	Pocket computers, electronic map-navigators, high-resolution TVs	Robots that can see, freeze-frame TVs, computers that recognize and use natural languages	Star Wars systems, personal robots, computers with human-like logic

Figure 1.1 The new superchips: what packing more power on a chip will bring.

Pervasive Influence

Computers are influencing every sector of the economy. Chips are cropping up in more and more products and production processes, while the personal computer has rapidly established itself as a standard working tool for employees almost everywhere. With the new all-electronic telephone exchanges, telecommunications networks are being transformed from simple voice transmission to data transmission and processing. Telecoms companies are becoming more like computer companies and vice versa. The telecommunications infrastructure now under construction, based upon fiber optics and augmented by satellite and cellular transmission, is probably as significant a development for society as the building of the railroads and the highway network.

In manufacturing industry, new techniques of factory automation made possible by microelectronics are bringing about a revolution in industrial production greater than anything seen for decades. Robots, computer-aided-design and manufacture (CAD/CAM) and flexible manufacturing systems (FMS) are creating the conditions for computer-integrated manufacturing (CIM), in which computers control fully integrated, automated factories. The completely unmanned 'Factory of the Future' is now a distinct possibility – although whether this is a good thing or a bad thing is open to question.

In offices, we are witnessing a dramatic shift from traditional paperwork to the electronic office. So many computer terminals, personal computers and word processors have been purchased and installed in US offices that soon there will be two terminals for every white-collar worker. Although the fully automated Office of the Future has been slow to arrive and the productivity gains from office automation are difficult to discern, all that spending on electronic gadgetry should have an effect in time, and the pressure on companies to cut white-collar labor costs is still enormous. Whether the electronic office will be a nicer place to work in is a different matter; a lot depends on how managements handle reorganization.

Retailing and the financial sector are being changed dramatically by information technology because the whole of commerce is based upon information. In a determined effort to cut their costs and to beat the competition, banks and retailers are investing heavily in computers and automation. Plastic money is rapidly replacing paper. On Wall Street, in Tokyo and in the City of London, financial deregulation coupled with the swift introduction of new computer systems have wrought a high-tech revolution of immense significance.

The high-tech revolution is also having a major impact in more traditional sectors like agriculture, mining, construction, transport and the professions. Down on the farm, for instance, we are seeing the development of robot tractors, robot fruit pickers, robot milking machines ("cowbots") and even robot sheep-shearers; computerized irrigation systems that use sensors to calculate water and fertilizer needs in different parts of a field; automated chicken houses, automated packaging stations, automated weed-killers and semi-automated rice combines. In addition, there are ultrasonic meat scanners and electronic ID tags for animals which record their life history.

Even in the "backward" construction industry, companies are using computer-aided design (CAD) and experimenting with knowledge-based systems and remote control devices. We are seeing the increased use of automated materials handling and the construction of "smart" office buildings fully wired and fitted with sensors to control heat and light. The automobile industry is busily installing microprocessor engine control, computerized anti-lock braking systems and "smart" suspension in the latest cars. Fiber optic wiring, new display devices and in-car navigation aids will soon become commonplace.

The world of marketing, too, is turning to high technology. Market researchers can now correlate the purchases recorded at supermarket checkouts with the socioeconomic status of individual purchasers in order to monitor the effectiveness of locally placed ads. Computer networks linking manufacturers electronically with suppliers and customers via retailers are being installed rapidly, especially in Japan. Videodisc systems are helping sales staff to sell everything from cars to cosmetics. Computerized ticketing systems are in use everywhere from the Metro in Paris to the Tokyo bus service. Electronic newsrooms are transforming newspaper production and TV program-making in London and around the world.

High-tech is helping revive the fortunes of traditional industries in Western countries previously laid waste by Far East and Third World competition. Textiles, embroidery, tailoring and, for example, wallpaper design and manufacture have been given a boost by computer systems. So-called "sunset" manufacturing industries like steel, rubber and consumer durables (such as washing machines) have prospered again, albeit often on a smaller scale, with the adoption of high-tech manufacturing techniques. Even the ailing Swiss watch industry has been turned around, thanks to the advanced production technology used to produce the Swatch and Le Clip – cheap, plastic, but fashionable electronic watches which have trounced the Japanese competition.

New Technologies and Applications

High technology makes possible new techniques such as three-dimensional (3-D) modelling and the simulation of the performance of as-yet-unmade products. For example, CAD/CAM systems allow engineers to design aircraft or vehicles on screen and to study and/or modify that design in 3-D. Aerodynamic simulation programs using the enormous power of supercomputers enable design engineers to see how an aircraft will perform without resorting to wind tunnel testing. Indeed, cars, boats, planes and other products can be tested to "destruction" before they are even made! Production lines can be "run" before a factory is built. Downtown redevelopment projects can be displayed and examined in 3-D before a single brick is laid. It may even be possible to "test" nuclear weapons without causing an explosion – which, I suppose, represents some kind of progress.

High technology has opened up new application areas and created new services like on-line databases (sometimes called "electronic publishing") and desktop or in-house electronic publishing systems. General information services and especially business information services have been successful, while a variety of legal, market research and political databases are now coming on-stream, some using a new kind of software called "relational database technology."

Retail stores, banks, brokerage houses and airline reservation offices are increasingly using on-line transaction processing (OLTP) to obtain instant information and to conduct transactions. In factories and warehouses, the growing popularity of "just-in-time" inventory systems makes such networks a hot new growth area.

All that so-called "desktop publishing" requires is a personal computer, page make-up software and a laser printer which does charts, graphics and headlines. Such a simple system can produce, for example, a modest newsletter for immediate distribution. Recent price falls have created a boom in demand for desktop publishing systems. Future improvements in text transfer will ensure that desktop publishing presents a major challenge to conventional print shops.

From desktop publishing, it is but a small step to more sophisticated in-house electronic publishing or computer-aided publishing (CAP), which is becoming increasingly popular. More and more companies are turning to "do-it-yourself" in order to cut the cost of producing technical and service manuals, legal documents, magazines, stationery, forms, business cards and fliers. They are finding that CAP gives a relatively rapid return on investment, and computerization further allows them to store material on disk, which makes updating easier.

In a lighter vein, we should note that the high-tech revolution has given us not only the hand-held computer for people like warehouse workers, and micro-TVs for in-car and in-flight entertainment, but also the wristwatch terminal or pager and the shirt-pocket telephone. There are now high-tech vending machines and high-tech time clocks, and a Florida firm is even selling high-tech birthday cards which contain a greetings program on a floppy disk.

For cars, there is now a high-tech breath analyzer which prevents drunk drivers from engaging a car's ignition and a high-tech black box journey recorder similar to that used in aircraft. For the home, there are, of course, high-tech bathrooms and high-tech kitchens full of food processors and intelligent washing machines. After eating, you can visit your high-tech dentist for computerized tooth replacement and perhaps bring the kids back a high-tech teddy bear.

Computing itself has not been standing still. Forty years since the first mainframes were switched on, it seems that they are on the way out: sales of personal computers, minis and superminis are growing, while those of mainframes are static or falling. "Downsizing" is now the name of the game in hardware as the price of IBM PC-compatibles or "clones" falls and further stimulates demand. Better architectures such as the reduced instruction set computer (RISC) will speed this process.

Software is gaining in importance relative to hardware: market researchers IDC say the world market for software will be up from $30 billion in 1984 to $110 billion by 1990, while hardware sales wil grow only from $60 billion in 1984 to $123 billion by 1990. Despite talk of a "software crisis" because of the time it takes programmers to churn out programs, the increased use of packaged software and now fourth-generation programming languages should help reduce the backlog. New network standards like OSI (open systems interconnection), MAP (manufacturing automation protocol) and TOP (technical office protocol) will also help out by reducing the need for specialized software.

Other developments to watch include voice processing and voice recognition systems, including "talkwriters," which promise to revolutionize the human–machine interface. Text processing using machine vision systems is also on the way. Perhaps most exciting of all, optical processing could become a reality with the development of optical computers that use blindingly fast light pulses instead of electrical signals.

Bell Labs have already built prototype optical chips, and optical switches have been developed for use in fiber optic telecommunications networks. Superfast optical computers could even supercede supercomputers for use in scientific and military applications, but the commercial optical computer is still a long way off. The potential, however, is enormous because of the speed of optical computing and the fact that lasers do not cause short-circuits when they cross.

In the short term, optical data storage on 12-inch disks or CD ROMs – the computer version of the audio compact disc (CD) – will be the first major application of optical processing. CD ROMs in particular could hit the big time as more and more companies endorse them. At the same time, CCDs (charge coupled devices) or imager chips are being widely adopted – most recently in the "filmless" camera, developed by Sony and first sold by Canon, which makes digital photography a reality.

Finally, developments in liquid crystal displays (LCDs) and electro-luminescent panels (ELs) are expected to come thick and fast. LCDs are already widely used in watches, calculators and portable personal computers, but in the next five years we should see their use in tiny TVs, wall-hung flat-panel TVs and even on advertisement hoardings.

The path of technological progress is not always smooth, however. Technological innovations interact with economic, social and ideological factors to influence the process of change. As Lynn M. Salerno points out in her book, *Computer Briefing* (John Wiley, New York, 1986), the high-tech revolution has taken a familiar path: a rosy scenario, painted by over-optimistic techno-boosters, has given way to a more sensible assessment as bugs develop in systems, users have trouble adapting their production processes, managers and staff find adjustments painful and economic recession plays havoc with investment plans.

Even so, the pace of change can be deceptive. Computers have been commercially available only for about half the average lifetime; the microprocessor chip was invented only in 1971, and mass-produced personal computers appeared on the scene only in 1977. It is only recently that ready-to-install robots, word processors, automated teller machines (ATMs) and so on have become available. In many ways it can be argued that the speed of the high-tech revolution is actually much faster than that of the Industrial Revolution, whose "revolutionary" nature has been questioned by historians anyway.

Yet, just as the speed of change is, by historical standards, extraordinary, so is the vastness of the landscape to be changed. Although high technology is penetrating every sector of the economy, huge areas of our social life remain relatively untouched by computers. Some have seen in California's Silicon Valley the seeds of a new High-Tech Society, but there is not much sign of High-Tech Man in most other parts of the globe. Also, High-Tech Society will not be without its problems, as I make clear below. But this has not stopped governments all over the world from trying to replicate the Silicon Valley phenomenon, a subject to which we now turn.

High-Tech Fever

World trade in high-tech goods is increasing by leaps and bounds: estimates vary and definitions are difficult, but the global market for information technology goods such as microchips, computers and telecommunications products will grow from about $200 billion in 1987 to about $600–$700 billion by the mid-1990s. In the USA the high-tech sector accounts for 3–6 percent of employees and 6–12 percent of GNP – and is growing fast. The same is broadly true of high-tech industries in Europe, Japan and the Far East.

As the high-tech pie has expanded, so has the international competition to get a bigger slice. Competition in high-tech is intensifying because the stakes are not only economic, but also political and strategic. Whoever dominates in high-tech will dominate in everything else. Battle-lines are being drawn as nations strive to keep pace in this international high-tech race. Every major country is taking steps to boost its domestic high-tech sector with a wide variety of support schemes. They are also competing fiercely to attract high-tech industries from elsewhere – usually the USA and Japan – with some very attractive packages on offer everywhere, from Europe and the Far East to the Caribbean. In addition, the Eastern bloc nations are striving desperately to keep up with the West, chiefly by stealing US technology, and the Chinese are also joining in. The whole world, it seems, has caught high-tech fever.

USA

In the USA, most organized efforts to boost high-tech have emanated from the private sector, especially in the form of joint research ventures. These include the Microelectronics and Computer Technology Corporation (MCC), founded by 12 major companies including Control Data, Dec, Honeywell, Motorola and Sperry in 1982 under the leadership of Admiral Bobby Ray Inman, former deputy director of the CIA (he quit in September 1986). MCC was later joined by other companies and established a base in Austin, Texas. It began work in 1984.

The Semiconductor Research Corporation (SRC), backed by a total of about 35 companies including IBM, Hewlett-Packard, Intel, RCA and some of the MCC sponsors, was also founded in 1982 and plumped for Research Triangle Park, North Carolina, as its base. The SRC's main role has been to commission basic research on microelectronics from key universities and to build collaborative work between academics and industry. The Pentagon's Defense Advanced Research Projects Agency (DARPA), Strategic Computing Plan (SCP) and Strategic Defense Initiative (SDI) are also pouring millions into microchip, supercomputer and AI research. A further $1.6 billion funding for the US semiconductor industry was proposed by a Defense Science Board (DSB) task force in December 1986.

The Small Business Innovation Research (SBIR) program, begun in 1982, requires federal agencies to earmark a higher proportion of their R&D funds for small high-tech companies. Small firms employ 60 percent of the US workforce; they have received more than 60 percent of patents issued; and they produce on average 2.5 times more innovations per employee than larger firms – yet in recent years they have been receiving less than 6 percent of federal R&D contracts, and large firms were 2.8 times as likely to receive an award as a small firm, according to a recent survey.

High-tech fever has hit Capitol Hill, largely in response to the Japanese challenge. Congressmen and senators have vied with each other to propose additions to the federal R&D budget and to suggest improvements in science and technical education in order to boost America's high-tech hopes. The Democrats favor a more interventionist role for the federal government, with proposals for central planning agencies and industrial banks, while the Republicans favor less government involvement: they believe that the government should concentrate on creating an economic environment in which high-tech ideas and companies can flourish. Despite these differences, there is general all-party support for research consortia like MCC and SRC –

although it is too early to say whether they will actually deliver the goods. Prior to 1980, only four states – Massachusetts, North Carolina, Connecticut and Florida – had programs to promote and attract high-tech industries. Even California, Texas and New York relied entirely on private universities and private venture capitalists to develop high-tech ideas. Today, no less than 38 states boast comprehensive high-tech development programs – and most other states are planning to follow suit. From the Eastern seaboard to the West Coast, from the Rust Bowl to the Sun Belt, everyone wants a piece of the high-tech action – especially the jobs, the tax revenues and the prestige that goes with it. When the MCC let it be known it was looking for a site for its new headquarters, 57 communities from 22 states made approaches, deluging the consortium with package deals and ever-increasing financial offers.

States are introducing tax incentives and putting up cash in the form of venture capital, sometimes taken from the pension funds of state employees. Several have built stylish, low-rent "incubator" facilities for high-tech start-ups, while many are developing "technology centers" or "science parks" in conjunction with nearby universities and colleges. Other states have established "networks" in order to bring together local investors and entrepreneurs, while virtually every state claims to be pouring resources into education and training, particularly the technical departments of colleges and universities.

Most states are indulging in the traditional arts of self-promotion and hustling. Following the success of Silicon Valley, civic boosters have given us Silicon Prairie (Texas), Silicon Mountain (Colorado), Silicon Desert (Arizona), Silicon Bayou (Louisiana), Silicon Tundra (Minnesota), Silicon Forest (Oregon) and Silican Valley East (Troy–Albany–Schenectady, New York). In pursuit of the Holy Grail of high-tech, urban planners from Atlanta and North Carolina to Pittsburgh and New Hampshire and Salt Lake City and Seattle are extolling the virtues of their areas, each claiming superior educational and research facilities, a better business climate and a higher quality of life.

"A lot of these programs are smoke and mirrors, without much substance," says Ira Magaziner, co-author of *Minding America's Business* (Harcourt, Brace, Jovanovich, New York, 1982). Many represent mere wishful thinking, or are packaged hype. It is also difficult to distinguish truly new measures from schemes that have been in operation for many years and have simply been repackaged in high-tech colors. But a more serious question arises as to whether states are simply playing "beggar thy neighbour" by stealing each other's companies. A 1984 Office of Technology Assessment Report (*Technology, Innovation and Regional Economic Development*) urged states to concentrate on encouraging home-grown high-tech innovation, rather than trying to attract it from outside. This would provide net improvements in the national economy rather than just shifting resources around.

Like the joint research ventures, it is too early to draw firm conclusions about the success of state efforts to attract high-tech. Many are not expected to pay off for a decade or more, and the benefits are hard to measure anyway. But there's little doubt that most are having *some* kind of positive effect. As Magaziner says, "It's easy to throw up your hands and glibly say the marketplace will take care of everything. But governments *do* make a difference, especially in accelerating the flow of ideas, people and technology between universities and businesses."

Britain

Similar arguments have raged in Europe, but the key difference is that European governments have long played a much more interventionist role. In the UK, for instance, the arrival of the microchip prompted a rash of official reports in 1978–9, which highlighted Britain's weaknesses in information technology (IT) and the generally backward state of much of British industry. As a direct response, the UK government introduced a variety of measures designed to encourage the domestic IT industry. Grants were made available for new investment, awareness campaigns were launched, education and training programs were beefed-up and, even more boldly, the government directly intervened to create new high-tech companies – one of which (the microchip manufacturer Inmos) survives today.

Despite the change from Labour to Conservative administration in 1979, most of the financial aid schemes were continued, although there was a shift of emphasis in some. Indeed, the Thatcher government declared 1982 to be "IT Year," and even created what was claimed to be the world's first IT minister. The exact nature of the various high-tech support schemes altered in succeeding years, usually to reflect changing UK priorities or changing technologies. A series of government-sponsored office automation pilot projects went the way of many grandiose Office-of-the-Future plans, while the Microelectronics Applications Project was judged a success in creating a greater national awareness of IT. But whether these earlier government aid schemes have really done anything concrete to stem the UK's relative long-term decline in information technology is debatable.

More promising, perhaps, is the £350 million Alvey Programme of research into fifth-generation computers, named after the British Telecom man who chaired the committee which proposed it and launched in 1983. Alvey projects sponsored in 1984–6 covered VLSI (very large-scale integration) chips, software engineering, expert systems and the human–machine interface. Alvey still has a while to run, and it is too early to say whether it has been a success.

Efforts at local level in the UK to encourage high-tech industry are in the hands of regional, county and city authorities. Among the most successful is the Scottish Development Agency (SDA) campaign to promote Silicon Glen, which is actually the central belt of Scotland from Greenock and Glasgow on the Clyde to Edinburgh on the Firth of Forth. Silicon Glen boasts many big names such as IBM, NEC, Hewlett-Packard, Motorola and National Semiconductor, and claims the largest concentraton of semiconductor plants in Europe. A number of major firms such as Wang, Dec and Apollo have also been attracted to the nearby University of Stirling. But while the SDA has enjoyed success in attracting foreign firms, there is concern in Scotland about the lack of indigenous high-tech industry growth and the levelling-off of job creation. There is also an underlying fear that US and Japanese firms could pull out and leave Silicon Glen just as easily as they came.

Three other high-tech concentrations in the UK owe their success less to agency planning and more to "natural" advantages such as good communications, educational and research facilities and high-quality residential areas. First, the Thames Valley, or M4 Corridor, has the greatest number of high-tech companies, covering as it does a large area from the outskirts of London all the way to Bristol and South Wales. Swindon, Newbury, Reading and Bracknell in the Corridor are Britain's boom towns, with high growth rates and low levels of unemployment. Second, the emerging M3/M27 Corridor has recently seen a rapid growth in high-tech companies. South Hampshire in particular is doing well, and this is probably related to IBM's decision to locate at Havant, Portsmouth, some years ago.

Third, the university city of Cambridge in East Anglia – or Silicon Fen, as it has been dubbed – has seen the creation of over 250 high-tech firms and as many as 20,000 jobs between 1979 and 1985. Although it built Britain's first science park, Cambridge's achievement has been largely unplanned and is clearly related to the excellence of the university's research facilities and the calibre of the people attracted to them.

In fact, following the success in the USA of the Stanford Research Park and Route 128, there was a mad rush by UK universities in the early 1980s to create science parks. Cambridge and Edinburgh pioneered, closely followed by Aston, Bradford, Brunel, Manchester and Warwick. A second wave financed by regional and local authorities in places like Warrington and Wrexham followed, together with some speculative property developments like Aztec West, Bristol, and the Solent Business Park in Hampshire. By 1986 over 20 science parks had been built, with another 20 or so planned.

Europe

European efforts to boost high-tech have been dominated by three pan-European projects: ESPRIT (European Strategic Programme for Research into Information Technology) and RACE (Research into Advanced Communications in Europe), both announced in 1982, and EUREKA (European Research Coordination Agency), formed in 1985. All three were set up in response to the falling market share of European companies in all categories of IT and the growing European trade deficit in IT goods – which marked a complete turnabout from the surplus enjoyed in the late 1970s.

ESPRIT, jointly financed by the EEC and industry, has underwritten basic research in five key areas targeted for rapid growth, such as Local Area Networks (LANs). The project will cost $1.3 billion over five years. RACE is on a much smaller scale, but is doing vital work on trying to develop common European telecommunications standards. The initiative for EUREKA came from the French, who saw it as an alternative to participation in the Reagan administration's Strategic Defense Initiative (SDI) program of research into "Star Wars" weapons. The UK and West Germany have now taken a major role in EUREKA. At a EUREKA summit in London in July 1986, 62 joint research projects worth $2.1 billion were approved by the 18 participating countries.

While these publicly funded joint research ventures have got underway, two notable private sector initiatives have also been launched. The Philips–Siemens Megaproject to develop the next generation of superchips has brought together the Dutch and the West Germans (and their governments, who are also providing subsidies). European Silicon Structures (ES2) is a pan-European start-up in the same business headed by France's Jean-Luc Grand-Clement and Britain's Robb Wilmot.

The French government continues to be a big IT spender – on computer hardware and software development, micros in schools programs, free videotex terminals, fiber optics, satellites and so on – but the French are having trouble selling their "Télématique" abroad, especially in the USA. More recently, the West Germans have shown signs of matching French spending as concern has grown about their weaknesses in computers and electronics.

Between 1980 and 1985, the European total of science parks increased from 10 to 47, with the biggest growth being recorded in West Germany, which had 18 compared to none in 1980. Most high-tech development in Europe's industrial powerhouse has been around Stuttgart in Baden-Württemberg and Munich, capital of Bavaria. In fact, with firms like Siemens, Texas Instruments and Motorola, Bavaria has become known as West Germany's "Siliziumtal" or Silicon Valley.

Meanwhile, the Republic of Ireland is attempting to make the transition from potatoes to chips, having achieved considerable success in attracting US firms with generous financial incentives. But Irish luck ran out with the failure of Gene Amdahl's ambitious Trilogy venture, which was to have had a manufacturing base in Eire, and the sudden closure of the Mostek plant in Dublin. Spain, now a member of the EEC, is hoping for greater success with "El Silicon Valley," its new promotional name for Madrid and Barcelona, where generous financial aid for US and Japanese firms is being offered.

Japan

The Japanese, of course, know all about high-tech fever – they practically invented it way back in 1971, when the Japanese government published its *Plan for an Information Society*. This, remember, was written pre-chip, but it didn't stop the Japanese from recognizing the key importance of computers, information and thus information technology in the years to come.

Japanese industrial policy has three key aspects, and all three have been brought into play in the case of high-tech. First, government and business work together to develop new strategies. The Japanese government does not intervene directly in the management of firms, nor does it run nationalized industries. Rather, it stimulates economic growth by supplementing market forces and anticipating major technological trends – thus the far-reaching "Plans" for the Information Society, the Information Network System (INS), Human Frontiers, Technopolis and so on. In this way, the government encourages the private sector to make the necessary adjustments. Japanese planning is thus a kind of programming for the future but in a competitive economy.

Second, the Japanese "target" industries for major export pushes and eventual domination. Heavy price-cutting and massive promotion are the main weapons here. By such methods, Japan has been enormously successful in picking off one industry after another, from motorcyles, machine tools and cars to cameras, TVs, VCRs and the whole range of consumer electronics. More recently, the successful targeting of semiconductors has led to considerable tension between Japan and

America, with US semiconductor firms alleging that Japan is "dumping" chips on the US market at unrealistic prices. It remains to be seen whether the July 1986 accord solves this particular problem.

Third, the Japanese identify key technologies for further research and development. Most Japanese R&D is applied and commercial in orientation, but increasingly more resources are being put into basic research. Since 1971, the Agency of Industrial Science and Technology (AIST), a branch of the Ministry of International Trade & Industry (MITI), has launched a series of research projects ranging from energy to ceramics to biotechnology. But of specific interest here are projects to develop high-speed computers, new types of chips, and better robots, databases, satellites and software.

The most significant technology project of all is the Fifth-Generation Computer Project, launched in 1981, which runs for ten years and is expected to produce a new generation of "intelligent" or knowledge-processing systems. Work under this project covers the areas of computer architecture, software engineering, expert systems, natural language processing and automatic translation. Skeptics question whether the Japanese will be able to achieve the breakthroughs they plan by 1991, but if they do succeed they will take over world leadership in artificial intelligence and the next generation of computers.

Of further interest is MITI's Technopolis Plan, conceived at national level in 1980 and implemented at local level starting in 1983. This plan designated 19 high-tech communities in Japan's hinterland which will become the focus of high-tech developments into the 1990s. High-tech firms are being steered toward the technopolises by a variety of financial carrots and sticks, thus creating the Japanese phenomenon of "U-turn workers" – skilled employees who have returned from the overcrowded cities to rural areas. As with many similar programs in the West, critics say that the Technopolis Plan is more marketing hype than substance. But it does seem to be working, especially in the case of Kumamoto on Japan's southern island of Kyushu, now dubbed Silicon Island.

The high-tech revolution is a reality, and it is proceeding apace. Every country in the world recognizes this, which is why they are trying feverishly to climb aboard the high-tech bandwagon before it's too late.

But how did this come about? Where did it all begin? The high-tech story is a tangled one, but perhaps it can best be told in the form of nine sub-plots: (1) the remarkable story of the computer and the magical microchip; (2) the unique history of the semiconductor industry's birthplace, Silicon Valley; (3) the fascinating tale of the far-reaching revolution in telecommunications; (4) the strange and sudden rise of the personal computer; (5, 6 and 7) the two-sided story of high-tech's impact so far in factories, offices and the world of money; (8) an account of the major problems likely to be faced by a high-tech society; and (9) a consideration of some of the frightening things that might happen to major world players like the USA and Europe in the very near future.

2 The Computer Revolution

Beginnings – How Chips are Made: the "Laws" of Microelectronics – New Types of Chip – New Ways of Making Chips – Supercomputers – The Fifth-Generation Project – Toward Artificial Intelligence (AI)

The world's first electronic digital computer, Colossus, was built in 1943 at Britain's secret wartime code-breaking center, Bletchley Park in Hertfordshire. Ten such machines were built, with the express purpose of reading German messages passed through their supposedly uncrackable code machine, Enigma. Throughout the rest of the war, the German high command continued to use Enigma, blissfully unaware that British intelligence was listening in. In such ways, the war was won for the Allies and Nazi Germany was destroyed. There could hardly have been a better demonstration of the value of computers.

Meanwhile, over at the Moore School of Electrical Engineering in the University of Pennsylvania, a Dr John Mauchly and a J. Presper Eckert were working on a machine that could do more than just crack codes. When finally switched on in 1946, their ENIAC (Electronic Numerical Integrator and Calculator) was capable of making the thousands of calculations per minute needed to assess the performance of new guns and missiles. One version was later to be used in the development of the hydrogen bomb.

ENIAC's 9-ft-high metal cabinets, bristling with dials, wires and indicator lights, weighed 30 tons and filled the space of a small gymnasium. It generated so much heat that industrial cooling fans were needed to prevent its circuitry from melting down. The machine contained no less than 70,000 resistors and 18,000 vacuum tubes, which failed on average at the rate of one every seven minutes. It cost over $2 million at today's prices and used so much electricity, it is said, that the lights of Philadelphia dimmed when ENIAC was turned on.

A handful of similar machines were constructed elsewhere in the USA and in Britain. But it seemed unlikely that many of these bulky, cumbersome and temperamental monsters would ever be built outside of the universities, military installations and government laboratories. Someone predicted that only four or five computers would be needed to fulfill the computing needs of the entire world, and in 1948 IBM itself decided, on the basis of market research, that there would never be enough demand to justify its entering the market for commercial computers.

Nevertheless, the first commercial computer, Univac 1 – another Mauchly–Eckert production – was installed at General Electric in 1954, and IBM soon changed its mind. Within a decade there were over 15,000 mainframe computers in the world. The 1960s and the early 1970s were the heyday of the mainframe and "data processing." Sales boomed and companies fell over themselves to install these new "electronic brains," the most common of which, the IBM 1401, cost the equivalent of just on $1 million.

Today, the same amount of computing power is contained in your average home computer costing perhaps $200. In fact, the very first microprocessor chip, the Intel 4004, introduced in 1971, was roughly equal in computing power to ENIAC. And by 1977 Intel founder Robert N. Noyce was able to claim, in the magazine *Scientific American*, that a typical microcomputer was much more powerful than ENIAC. Moreover, "It is twenty times faster, has a larger memory, is thousands of times more reliable, consumes the power of a light bulb rather than that of a locomotive, occupies 1/30,000 the volume and costs 1/10,000 as much."

Put another way, if the automobile and airplane businesses had developed like the computer business, a Rolls Royce would cost $2.75 and run for 3 million miles on one gallon of gas. And a Boeing 767 would cost just $500 and would circle the globe in 20 minutes on five gallons of gas.

This, then, is the shape and size of the revolution in computing brought about by a remarkable innovation, microelectronics.

Beginnings

The microelectronics revolution really began at about the same time as ENIAC was unveiled, although nobody realized it at the time. It was made possible by three key inventions: the transistor, the planar process and the integrated circuit.

In 1947, three scientists working on telecommunications research at Bell Laboratories in the USA first discovered the transistor effect. While searching for switches and amplifiers to replace mechanical relays and the

valves that so troubled ENIAC, J. Bardeen, W. H. Brattain and William Shockley came up with the point contact transistor, a small piece of germanium with wire contacts which functioned as a simple, if weak, amplifier. Its successor, the junction transistor, was also made of germanium, though this was later changed to pure crystals of silicon, the main ingredient of beach sand.

By the mid-1950s there was a score of firms producing different types of transistors. Many of these firms were spin-offs from Bell and other large companies. As Ernest Braun and Stuart Macdonald point out in their book, *Revolution in Miniature* (Cambridge University Press, New York, 1982), a new breed of person had emerged: the scientific entrepreneur. By 1957 the new firms had captured 64 percent of the transistor market. What is more, the really crucial innovations were emerging from these small firms, which had attracted the brightest stars out of the established corporations with the lure of fame – and fortune.

And so it was that Fairchild Semiconductor, a company started only in 1957, produced the first transistor using the planar technique in 1959. The planar technique, invented by Jean Hoerni, consists of three processes. First, a slice or "wafer" of silicon is oxidized and then coated with a photosensitive material, a photoresist. Second, a pattern is photographed onto the resist, which becomes vulnerable to certain chemicals when exposed to light. The photographed pattern is etched through the resist and the oxide underneath. Third, the resist is washed off and impurities are allowed to diffuse into the exposed parts of the wafer, while the rest is protected by the remaining oxide layer. The overall process can be repeated and intricate patterns of conducting layers built up.

When transistors produced by the planar process hit the market, there were already 84 firms operating in what had become known as the semiconductor industry, which was then concentrated around Route 128 near Boston, Massachusetts, and Stanford University in California. Large military purchases of transistors for communications equipment boosted the business to such an extent that by 1963 nearly half of all transistor sales were to the US government. Semiconductor devices were also finding their way into transistor radios, hearing aids, television cameras and, of course, computers. But no one as yet had realized that semiconductor devices themselves would eventually *become* computers.

The next great step came with the invention of the integrated circuit, generally credited to Jack Kilby at Texas Instruments and Robert N. Noyce, then at Fairchild. Thanks to the planar process, it soon became apparent that whole circuits consisting of many transistors could be incorporated into a single silicon chip. Again, military orders helped firms to cover research costs and get their devices into mass production. Mass production meant that prices dropped and sales increased: during the 1960s shipments of transistors rose three times and integrated circuits 40 times. At the same time, the average price of an integrated circuit fell from about $30 to just $1 and the average number of components in a circuit rose from 24 to 64.

Each year, scientists packed more and more components onto a chip. Since making a complex chip cost only a little more than making a simple one, with each increase in complexity came a corresponding decrease in price per function. The incentive was there to design more sophisticated devices.

In addition, integrated circuits were neat and cheap, but they could perform only limited tasks. They had to be soldered into a rigid pattern on a printed circuit board – in other words, they had to be given strict instructions on what to do. The central processing units (CPUs) of large computers contained hundreds or thousands of integrated circuits. They were even less flexible than good old ENIAC's masses of wires, which could at least be moved around.

Enter the microprocessor. In 1969, a young engineer at a small firm called Intel, Marcian "Ted" Hoff, found himself making chips for an electronic calculator designed by a Japanese firm, Busicom. Hoff saw a way of improving on the Japanese design, and in 1971 he succeeded in putting the entire central processing unit of a computer on a single chip. This was made possible by the great advances in chip production, which were enabling more and more complex circuits to be placed on smaller and smaller surfaces.

Hoff's chip became known as a microprocessor. A microprocessor does not just react in fixed, pre-programmed ways like an integrated circuit. Its logic or response can be altered. This was a tremendous breakthrough, because by adding two memory chips – one to move data in and out of the CPU, the other to provide the program to drive the CPU – Hoff now had in his hand a rudimentary general purpose computer that could do far more than run a hand-held calculator: it could be used to control a set of traffic lights, an oven or washing machine, an elevator, or even a factory production line – the commercial potential was enormous.

Asked for his reaction to Hoff's invention, J. Presper Eckert said that, if anyone had suggested the idea to him in the days of ENIAC, he would have described it as "outlandish." When, in the late 1960s, Robert Noyce had predicted at a conference the coming of the computer-on-a-chip, delegates were aghast: "Gee," exclaimed one, "I certainly wouldn't want to lose my whole computer through a crack in the floor." Noyce told him not to worry: "You have it all wrong, because you'll have a hundred more sitting on your desk, so it won't matter if you lose one."

To summarize, modern chip technology began with the invention of the transistor in 1947 and broadly went through four stages of development. (1) In the 1950s single circuits were mounted on boards. (2) In the 1960s up to 100 circuits were built right into each board, and this meant that large-scale computing could be had for the first time for less than $1 million. (2) In the 1970s the boards got smaller and became chips. Each year, engineers packed more circuits into the same size of chip, and this meant that powerful computers could be purchased for less than $100,000. (4) In the early 1980s, hundreds of thousands of circuits could be packed into a chip and good computers could be brought for less than $10,000 – rudimentary ones cost as little as $100.

In the late 1980s, we are entering the 1 micron era, and in the 1990s we will enter the sub-micron era. (A micron, or micrometer, is 10^{-6} of a meter or about 1/25,000 of an inch, whereas the width of an electrical connection in an early 1980s chip was 4–6 microns.) After the year 2000, almost anything might happen.

Trends in miniaturization may be illustrated by means of the street map analogy (see figure 2.1). Roughly speaking, in the 1950s information about just one street could be placed on a chip or board; in the 1960s the

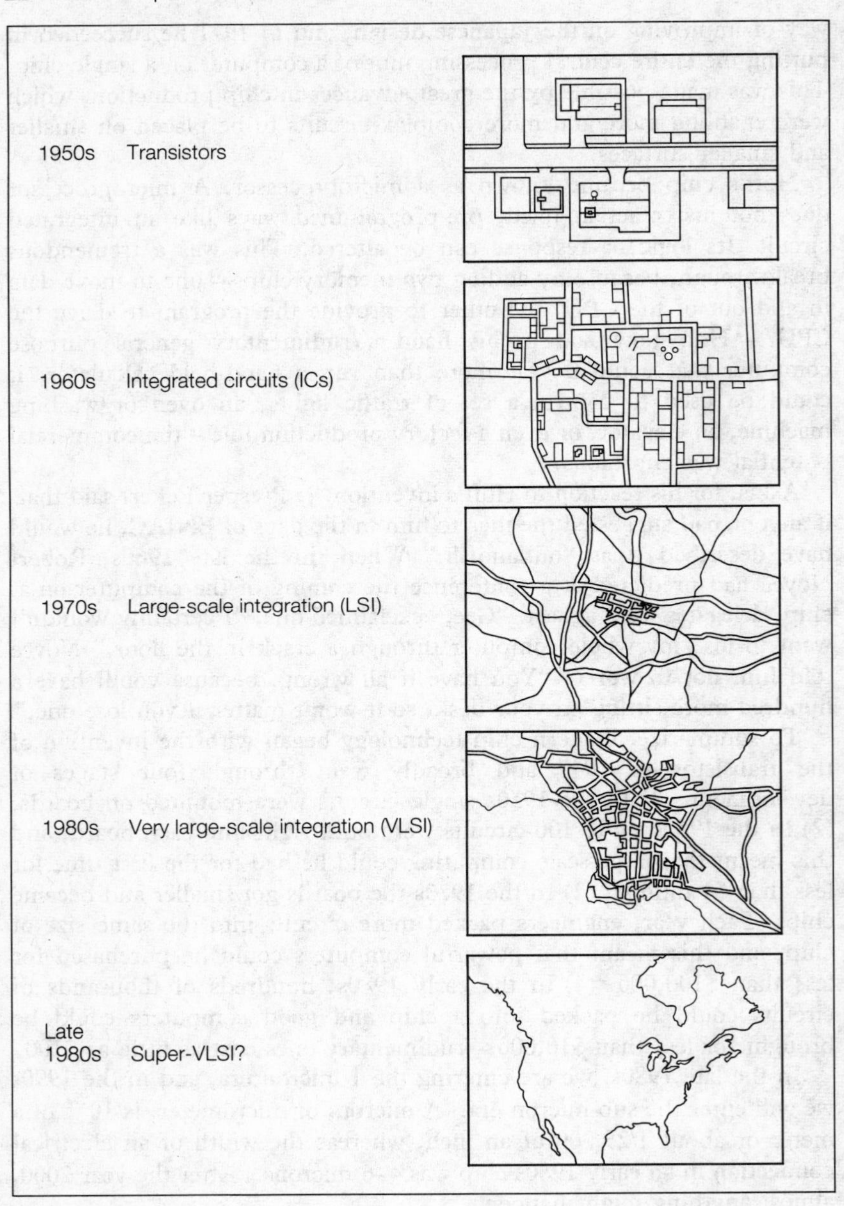

Figure 2.1 Trends in miniaturization.

street map of a small town could be placed on the same size of device; in the 1970s a chip could contain the street map of a smallish city, and in the mid-1980s the street map of the entire Greater Los Angeles area could be placed on a chip. Sometime in the late 1980s it will be possible to put a street map of the entire North American continent on just one tiny chip. In the 1990s, it could be the whole world.

How Chips are Made: the "Laws" of Microelectronics

Today's microchip typically contains tens of thousands or hundreds of thousands of circuits packed into an area the size of a pea. During the manufacturing process, each chip is broken off a 4 in. "wafer" of silicon which might contain 200 or so identical chips, and each wafer might be part of a batch of hundreds of similar wafers. Defective chips are common, but the yield of usable chips rises as firms learn more about the manufacture of each new chip (the so-called "learning curve"). A higher yield of good chips obviously means lower prices and higher sales, so in order to understand the remarkable economics of microelectronics it is helpful to understand something of the chip fabrication process.

A microchip begins life in the mind of a circuit designer, who will create the intricate layout of the new device with the help of computers, visual display terminals (VDTs) and light pens (see figure 2.2). In addition to aiding the design, computers are also used to simulate the operation of the circuit to see if it will actually work. When the design is complete, the computer commits it to its memory. From that description in the memory, a complete set of photomasks representing each layer of the circuit is prepared. The masks are taken to a wafer fabrication plant, where the workers wear protective clothing and the air is constantly filtered in order to eliminate dust particles, which can ruin a chip.

The fabrication process begins when raw silicon is reduced from its oxide, common beach sand. The silicon is purified and then heated to melting point in a crucible. From this crucible a cylinder of 4 in. diameter silicon is drawn, and this is sliced into 0.5 mm thick wafers using a high-speed diamond saw. The wafers are then loaded into slots on a quartz "boat" and pushed into a furnace, so that a layer of silicon dioxide forms on the surface. Silicon dioxide is an insulator and can act as a mask for the selective introduction of dopants – crucial in the chip fabrication process.

The chip's circuit is built up layer by layer, each layer receiving a pattern from the photomask prescribed in the design. This process, known as photolithography (see figure 2.3), has been described by

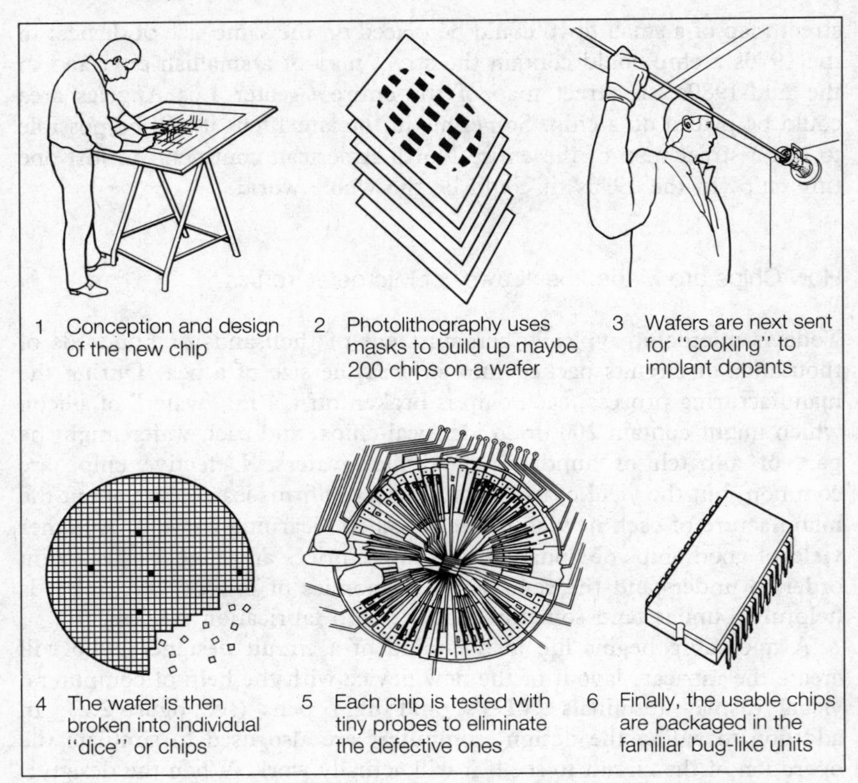

Figure 2.2 How chips are made: stages in the fabrication process.

William G. Oldham as "the key to microelectronic technology." An oxidized chip is first coated with a photoresist, whose solubility to certain solvents is greatly affected by exposure to ultraviolet radiation. After exposure through an appropriate photomask, the wafer containing the chips is alternatively washed in acid solutions and heated, as the intricate circuit is painstakingly constructed.

Active components in the circuit such as transistors are formed by the introduction of impurities or "dopants", to create negative and positive areas within the chip. Two techniques are used for selectively introducing dopants: diffusion and ion implantation. Diffusion occurs when the chips are heated to a high temperature in the presence of impurities such as boron and phosphorous, which move through the silicon where it is unprotected. Ion implantation is a means of introducing impurities at room temperatures by ionizing the atoms of dopants and accelerating them into selected parts of the chip. Transistors

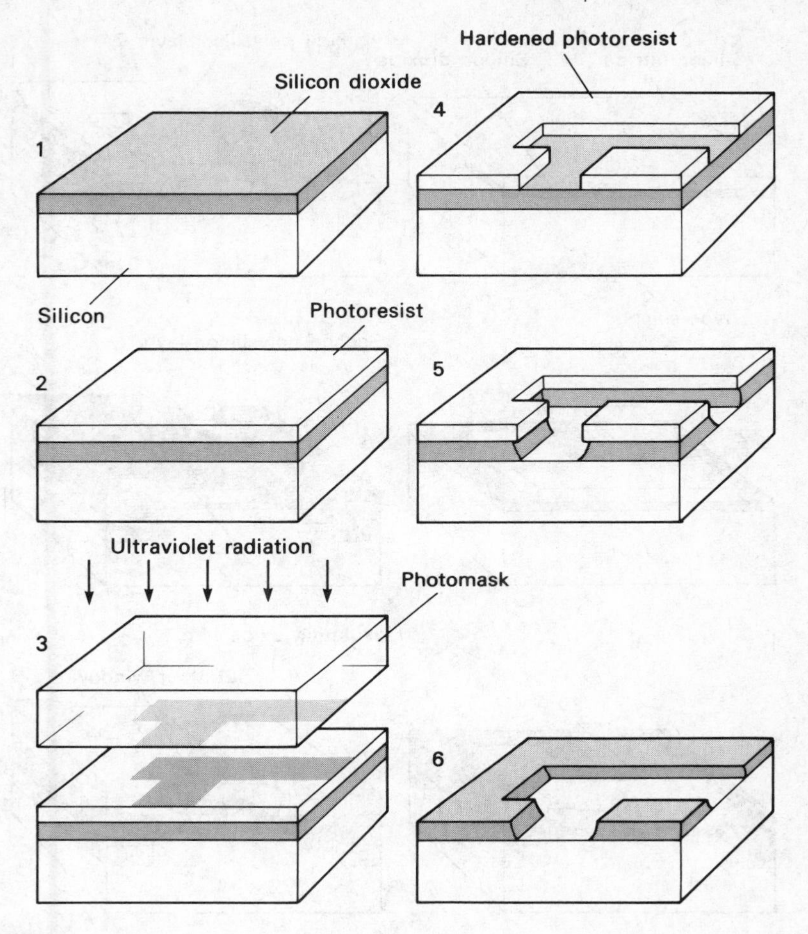

Figure 2.3 The process of photolithography.
(Source: "The Fabrication of Microelectronic Circuits" by William G. Oldham. Copyright © 1977 by Scientific American Inc., 1977. All rights reserved.)

formed in this way are called metal-oxide-semiconductor (MOS) devices, and this method of creating whole circuits on a chip is sometimes referred to as MOS technology (see figure 2.4).

Ion implantation techniques have improved greatly in recent years, enabling engineers to cram more and more transistors onto a single chip. In September 1985 Applied Materials Inc. announced a new range of sophisticated ion implantation machines first conceived by a British firm, Lintott Engineering of Horsham, Sussex, which Applied acquired in 1979. The major players in the ion implantation business in 1986 were

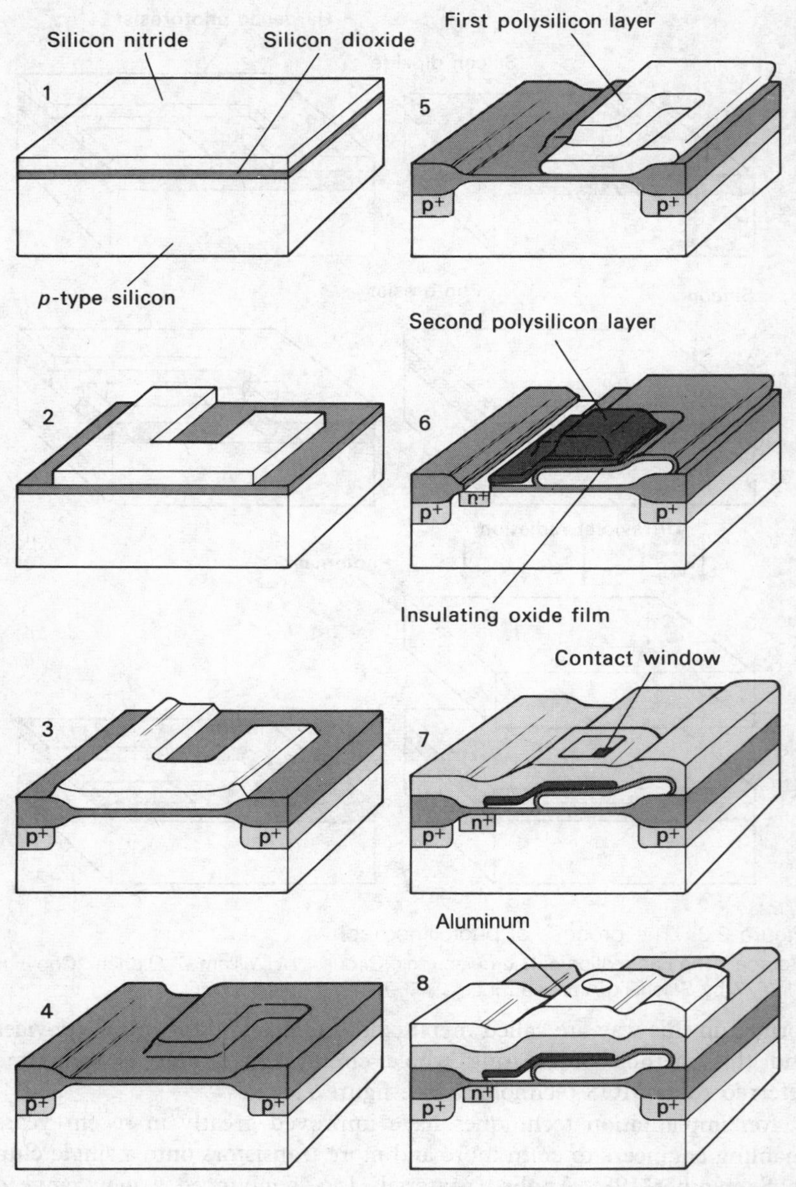

Figure 2.4 The complete fabrication sequence (of a two-level, *n*-channel polysilicon-gate MOS circuit element).
(Source: "The Fabrication of Microelectronic Circuits" by William G. Oldham.)

Extrion and Nova, both of Massachusetts, and VG Semicon, of East Grinstead, Sussex, England.

When the circuit is complete, it is sealed with an insulating film formed with the help of silane gas or silicon nitride and the new chip is electrically tested with probes on a computerized testing machine. Defective chips are marked and discarded when the "dice" are broken off the wafer and sent for packaging into the familiar units which resemble a centipede – a process that usually takes place in a cheap-labor country.

The unique nature of the chip fabrication process led Intel co-founder, Gordon E. Moore, to coin "Moore's Law." Back in 1964, Moore (then at Fairchild) noted that the number of circuit elements on a chip had been doubling every year. He suggested that complexity, or the industry's "experience," would continue to double every year – and, roughly speaking, it has. The result has been a persistent decline in the cost per electronic function as firms progress along the "learning curve" and produce an ever-higher yield of good chips. In fact, integrated circuit costs have decreased by an average 30 percent per year since 1973.

The reduction in the size of circuit elements not only reduces costs but also makes for a better chip. Because delay times are directly proportional to the dimensions of circuit elements, a chip actually becomes faster as it becomes smaller. It also generates less heat and uses less energy and is therefore more reliable and cheaper to run.

A good illustration of these "laws" is provided by the history of the pocket calculator, a product based entirely on a chip. Introduced in 1972, the first simple mass-produced electronic calculators cost around $250. Firms rushed to produce them, each progressing along the learning curve and using ever-more complex chips. The result was a dramatic decrease in price, from $250 to $100 to $50 to $20, so that by about 1977 a more sophisticated, more reliable and cheaper-to-run calculator could be had for less than $10. Only a few firms were left in the business, but the price reductions had opened up a vast new market for calculators.

Other industries have similar experience curves, but only microelectronics can boast a doubling of complexity, a doubling of output and a 30 percent reduction in prices every year. In such an industry, therefore, the motivation to do research and development in order to stay one jump ahead of the rest is overwhelming. A year's lead in introducing a new product gives a company a huge cost advantage. Laggardliness is punished by total failure. The development of the microelectronics industry has thus required a great number of entrepreneurs willing to take a gamble on new devices, ready supplies of venture capital, and a business climate conducive to entrepreneurial innovation.

There is no stopping the microelectronics revolution. The substitution of chips for other control devices leads to savings in labor time and materials. Chips require less maintenance and can be simply replaced if they do go wrong. Chips take up less space, so they are leading to great improvements in existing products as well as creating entirely new products. Chips are displacing other modes of control in production processes to such an extent that there are few areas of industry or commerce left untouched by microelectronics. That is why the word "pervasive" (along with "convergence") crops up so often in discussions of the impact of the computer revolution.

Pervasiveness is the hallmark of a new "core" technology, like steam power or electricity. Just as the first industrial revolution greatly extended the power of human muscles, so the new industrial revolution is extending the power of our nervous system and the human brain.

New Types of Chip

Chips are getting smaller, denser, cheaper and better. In the years since Ted Hoff first invented the microprocessor, it has evolved through four generations: the 4-bit, 8-bit, 16-bit and now 32-bit models. The very latest 32-bit microprocessors, for example, can process information in groups of 32 binary digits ("ones" or "zeros") at a time. This means that chips have now broken the 1 MIPS barrier – they can execute no less than 1 million instructions per second.

At the same time, the storage density of microchip memories has leapt on average four times every three years or so. Memory chips are called either ROMs (for read-only memory, where information in the chip cannot be altered), RAMs (for random access memory, denoting the abililty to read each memory cell at will) or D-RAMs (the "D" stands for "Dynamic," referring to the need to constantly recharge the memory to keep the data from vanishing). Their storage capacity is measured in "K", roughly representing 1000 memory cells (the actual figure is 1024).

Thus, the 1K RAM introduced by Intel in 1970 became the 4K RAM of 1973 and was in turn superceded by the 16K RAM introduced in 1976. IBM and the Japanese firm Fujitsu both lay claim to being the first to mass-produce the 64K RAM in 1979, and by 1982 examples of 256K RAMs were appearing. Investment costs had risen from $25 million for a 16K production line to $100 for the 64K RAM and a prohibitive $200 million for the 256K RAM. But meanwhile the price obtainable for each chip continued to fall: a 64K RAM, for example, sold for $15 in 1981, $6 in 1982 and only $3 in 1983 (see figure 2.5).

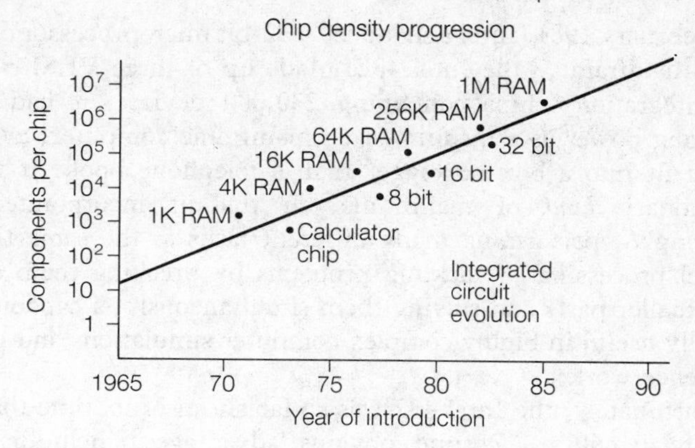

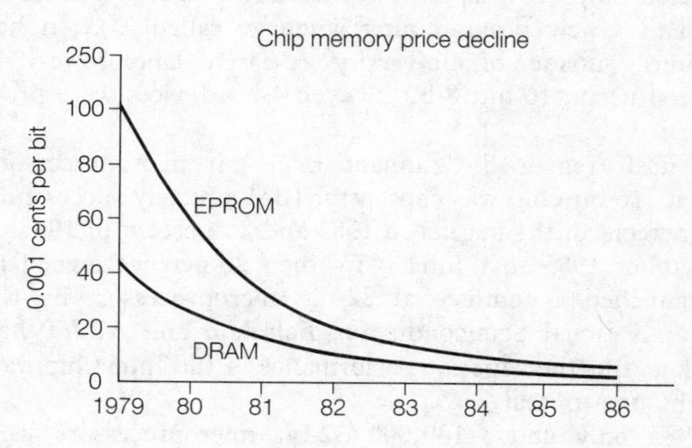

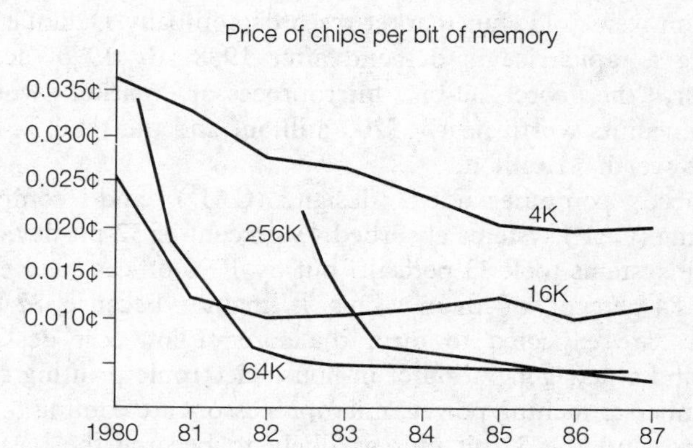

Figure 2.5 Denser means cheaper: the price of computing power declines as chips get better.
(Source: *Financial Times*)

In February 1981, Intel announced a 32-bit microprocessor or so-called "micro-mainframe," the Intel 432. Made up of three VLSI (very large-scale integration) chips containing 230,000 devices, it had the same computing power as a medium-sized mainframe computer, even though it could fit into a box no bigger than a telephone book. It was also a revolutionary *kind* of mainframe, in that it incorporated "multi-processing" – performing many different tasks at the same time – and "parallel processing" – tackling problems by breaking them down into many smaller parts and solving them simultaneously, a capability that is especially useful in highly complex computer simulations and in artificial intelligence work.

Unfortunately, the Intel 432 was so far ahead of its time that it was a commercial failure. Despite obvious advantages, including software savings and a new programming language called ADA, it had limited applications outside of university research laboratories. For most commercial users, 16-bit, 8-bit or even 4-bit devices were proving quite adequate.

Intel itself remained dominant in 16-bit microprocessors, largely because its 16-bit chip was chosen for IBM's hugely successful PC. Intel won 83 percent of the market in 1984 and 85 percent in 1985. It was not until October 1985 that Intel – by then 20 percent owned by IBM – finally launched a commercial 32-bit microprocessor, by which time Motorola, National Semiconductor, Fairchild and AT&T had entered the market. But the superior performance of the Intel chip meant it was clearly the one to beat.

In 1985 only about 100,000 32-bit microprocessors worth about $17 million were sold, but market research company Dataquest Inc. was predicting a rapid rise in demand after 1988. By 1990, according to Dataquest, the core 32-bit microprocessor market would reach 4.7 million units worth nearly $200 million, and the total 32-bit market would be worth $1 billion.

Advanced computer-aided design (CAD) and computer-aided engineering (CAE) systems absorbed 43 percent of 32-bit devices in 1985 and office systems took 33 percent, but by 1990 offices were expected to swallow 84 percent of them. This is mainly because 32-bit micro-processors are expected to form the core of low-cost desktop work-stations and a new generation of in-house electronic printing systems. In their number-crunching power, microprocessors are coming to rival large computers, and the 32-bit chip is likely to become the basic building block of the computer industry up to the turn of the century.

Also in October 1985, the British company Inmos announced a 32-bit device that amounted to a complete computer on a chip. Called a "transputer," this innovative microprocessor incorporates processing, memory and communications elements. The use of parallel processing and RISC (reduced instruction set computer) circuitry means that the transputer can achieve high processing speeds – between four and ten times faster than other 32-bit chips.

In theory, a string of transputers could achieve the speed and processing power of one of today's supercomputers costing $15 million. Inmos claims that the transputer is naturally suited to new applications such as computer graphics, vision systems, speech recognition and artificial intelligence. The future of Inmos – originally set up with UK government money – depends very much on whether the transputer is accepted by computer designers. The early omens were good: in March 1986, Floating Point Systems of Beaverton, Oregon, announced that the Inmos transputer was to star in its new supercomputer; more major orders came in during 1986, and both Matsushita of Japan and Siemens of West Germany expressed an interest in buying a stake in the once-languishing company.

Back in February 1982, Hewlett-Packard had announced a 32-bit microprocessor combined with a 128K D-RAM in a set of six chips equal in power and performance to a mid-range mainframe. The D-RAM alone contained 660,000 devices, one and a half times the number of parts in a jumbo jet – the most ever built into a single chip of silicon. It was produced using 1 micron technology and conventional production techniques, and the first commercial shipments followed in October 1983. At the same Solid State Circuits conference in San Francisco, IBM unveiled a 288K RAM, produced using electron beam lithography (see next section), which was smaller, faster and cheaper than anything seen before.

By early 1983, the battle for the 256K RAM market began in earnest as US manufacturers were spurred into action in the knowledge that Japanese firms had captured 70 percent of the 64K RAM market. As the production of 256K RAMs accelerated, prices fell from $110 in 1983 to less than $4 in 1985! A 256K RAM has 262,144 memory cells (256 × 1024) and in all contains some 600,000 devices on a single chip, which is placed in the same package used to enclose a 1K RAM chip. A single 256K RAM chip can store 10,000 telephone numbers, compared with 2500 in a 64K RAM chip. But it also provides *faster* access to data – below the 100 nanosecond barrier (a nanosecond is a billionth of a second or 10^{-9} of a second), compared with 150–250 nanoseconds in most 64K RAMs.

In September 1983, IBM announced a 512K RAM, said it was building a 1 million bit or megabit chip for internal use (that's four times the capacity of a 256K RAM chip) and claimed that an 8-million bit chip (that's 32 times the number of devices on a 256K chip) was possible. A year later, AT&T entered the megabit chip race, which by then included the Japanese firms NEC, Fujitsu and Hitachi.

"The megabit chip," says Vic Vyssotsky, executive director of Bell Lab's information services division, "is to the 64K RAM – the standard memory chip used in most of today's computers – what the jet engine was to the propeller." Meanwhile, in January 1985, South Korea became only the third country in the world to manufacture 256K RAMs when Samsung announced its device – by which time Japanese firms had 90 percent of the 256K RAM market.

With the arrival of megabit RAMs, microelectronics will soon be entering the sub-micron era. Megabit, or 1-million-bit, chips have circuit lines 1 micron thick. Four megabit chips, such as the three-dimensional one announced by Matsushita in June 1985, have circuit lines ½ micron thick. Gigabit chips – with 1 *billion* transistors on a chip – may be possible by the end of the century. Thus, VLSI (with 100,000 to 1 million transistors) is giving way to ULSI (ultra large-scale integration).

Megabit chips, or so-called "superchips," will soon be able to do the work of today's mainframes. By the mid-1990s they could contain more power than today's supercomputers, making them ideal for applications in artificial intelligence, vision systems and "Star Wars" weapons.

Megabit chips and the pursuit of gigabit chips are also spurring a revolution in chip-making itself. The process of making chips is becoming more capital-intensive, more automated – and much more expensive. In the new automated chip production facilities, the air has to be so clean that humans are best kept outside – so soon chips, in effect, will be making chips. The high investment costs are also having major implications for US–Japanese competition, while the Europeans are trying desperately to stay in the superchip race via the Philips-Siemens "Megaproject," ES2 and the proposed Thomson-SGS venture.

Two important types of chip to emerge in recent years are the EEPROM (or EEROM) and the CCD, or imager chip. An EEPROM is a non-volatile memory chip upon which old instructions can be removed instantaneously and new ones implanted. It stands for "electrically erasable, programmable read-only memory." RAMs and D-RAMs suffer instant and total memory loss when the power is turned off, whereas EEPROMs have a permanent memory (until reprogrammed) and are also very fast. They are thus ideal for use in point-of-sale terminals, for example, where information such as prices need to be updated, or in simplified telephone dialling machines, where there is a need for the machine to remember all it has been taught.

The CCD (charge-coupled device) or imager chip is a chip that can "see" 80 times more efficiently than photographic film. Invented in 1969 at Bell Labs by George Smith and Willard Boyce, CCDs have replaced bulky vacuum tubes known as vidicons and made possible a new generation of compact, solid-state video cameras. These are smaller, lighter and stronger than conventional TV cameras and are being used widely on production lines for super-accurate measuring, for directing robots and especially for quality control: CCDs are capable of instantaneously detecting the smallest of defects, even in chips themselves! They can find flaws the human eye could never see. For the consumer market, in November 1985 Sony launched its Handycam, an 8 mm solid-state video camera the size of a largish paperback book.

The invention of chip-based "vision systems" and sensing devices has opened up a fast-moving area of new technology in which CCDs are linked with transducers or analog microprocessors, which convert continuous flows of information into electrical signals for digital processing and then back again. Already, "electronic eyes" are being used in scanning systems in retail stores, in the optical character recognition (OCR) systems being introduced into post office sorting areas, and in the guidance systems of cruise missiles. "Seeing," "feeling" and "smelling" chips will make possible the construction of more sophisticated robots and new weapons systems, and will help in the development of artificial intelligence.

In perhaps the most spectacular demonstration of their value, CCDs were used by astronomers in late 1985 to detect the arrival of Halley's Comet months before it could be seen through telescopes and to chart its course across the heavens in early 1986. According to the magazine *New Scientist* (March 6, 1986), larger CCDs will undoubtedly become the standard detectors in telescopes of the future: "Astronomers have virtually achieved the quest for their Holy Grail – a perfect detector of light."

While CCDs are transforming astronomy, another new type of chip could bring artificial intelligence down to earth. The so-called "smart" chip has an "expert system" – the sum total of knowledge and rules of thumb that represent the highest level of expertise in a given field like oil drilling or medical diagnosis – actually embedded in it.

First developed at Bell Labs by two Japanese-born researchers, Masaki Togai and Hiroyuki Watanabe, the "smart" chip uses so-called fuzzy logic, which is a set of rules that deal with imprecise data by mimicking common-sense reasoning. It can therefore read between the lines, so to speak – saving time and processing power – and can cope with the gray areas of life, not just the black and white ones.

But in the short-term, so-called "smart power" devices are likely to be more important because they have the potential to transform automobiles, factories and consumer appliances. "Smart power" devices combine logic and electric power controls on a single chip. Such chips could replace bulky discrete components and the masses of wiring used in cars, robots and refrigerators.

Electric motors controlled by "smart power" chips would be able to attain variable speeds, thus increasing the efficiency (and reliability) of air conditioners, washing machines and other kitchen appliances. Existing stop/start actions wear out motors and waste energy. "Smart power" chips are already being used in computer printers and telephone circuits, but some say their real potential lies in such areas as lighting control and entirely new technologies like flat-panel displays. Only cost and conservatism among electronics engineers are holding things back.

So-called "micro-mechanical devices" are tiny valves, nozzles, sensors and other mechanical systems, made of silicon. Using modified semiconductor production equipment, the new discipline of micromechanics enables engineers to craft tiny, low-cost three-dimensional devices that will serve as the eyes and ears of computers. The possibilities include sensors to measure the pressure inside a blood vessel, components of an artificial ear for the deaf and microscopic switchboards for routing optical signals.

Another interesting area is that of wafer-scale integration, or so-called "monster" chips. Here, the general idea is that the wafer in the chip-making process is kept intact instead of being broken up into individual chips. The resulting giant chip in theory would perform faster and better than a lot of ordinary chips joined together. The dream of wafer-scale integration has already claimed one victim – Trilogy Systems, which abandoned the effort in 1984 after spending $230 million.

Yet others, in particular Mosaic Systems of Michigan, Inova Microelectronics Corporation of California and Britain's Sinclair Research – not to speak of most major semiconductor and telecoms companies – are taking up the "monster" chip challenge. Excessive heat, defective chip circuits and the sheer scale of the design task are just some of the problems wafer-scale chip-makers have run up against. But if these problems can be overcome, such giant chips will rival mainframe computers in computing power and will open up all kinds of applications possibilities in industry, defense and medicine.

Finally, in early 1985 news leaked out that Bell Labs were working on a so-called "ballistic" transistor, which would switch at the incredible speed of 10 femtoseconds (a femtosecond is one-quadrillionth or 10^{-15} of a second). Later, it emerged that Gould Inc. of the USA and Fujitsu of Japan, among others, were also in the race. Such speeds would be possible by making a transistor so small (like about 0.01 of a micron) that an electron could shoot straight through it without bumping into any other electrons or atoms in the semiconductor material. The "ballistic" effect is similar to a jet passing through the sound barrier.

If ballistic transistors really do emerge out of current research in the USA, Japan and France into so-called HEMTs (high electron mobility transistors) and RHETs (resonant tunneling hot electron transistors), they could open up a whole new generation of superfast computers, greatly improved satellite communications and advanced radar systems.

New Ways of Making Chips

Chip-manufacturing technology doesn't stand still. New *production processes*, such as electron beam, focused ion beam and X-ray lithography, are being developed, because 660,000 devices on a single chip seems to be about the upper limit using conventional photolithography. New *manufacturing technologies*, such as CMOS (complementary metal-oxide semiconductor), are becoming more popular; and new *materials*, such as gallium arsenide, are increasingly being used to replace silicon itself as the basis of chips.

Electron beam lithography enables chip fabricators to pack many more devices onto a chip than was hitherto possible. Instead of using a photomask, an electron beam guided by a computer actually traces the intricate patterns of a chip's circuit. Because electrons have much smaller wavelengths than light, they are capable of drawing much finer lines – as small as maybe 8 nanometers (a nanometer is 1/1000th of a micron or micrometer), compared with the 500 or so nanometers achievable through conventional photolithography.

With focused ion beam lithography (FIBS), ions instead of electrons are accelerated through a series of focusing lenses onto the chip to etch out the circuit lines. This makes for a more automated production process, but the actual FIBS procedure is slower than electron beam lithography. FIBS is unlikely to be adopted for the mass production of standard chips, but it may be used to produce custom chips or "gate arrays" (see below), for which there is a growing market.

Researchers are also investigating the possibility of using X-rays in the production of chips. As in photolithography, a mask would be used, but the wafer of silicon would be coated with an X-ray-sensitive rather than a light-sensitive resist, into which the circuit pattern is etched. Problems include cost, slowness and the difficulties of constructing a mask that stands up to regular use. In addition, lasers are figuring seriously in the plans of those working on the next generation of chips. Potentially, they could be much cheaper to use than expensive electron beam machines.

Plasma etching is of growing importance in the production of the new generation of superchips. This involves the slicing away of microscopic slivers of material to open up gaps in the thin layers of semiconductor from which integrated circuits are made. A plasma is an array of highly reactive ions in a gaseous form. In plasma etching, the ions cut not only downwards but also sideways to make the required hole. The next stage is reactive ion etching, a form of plasma etching in which the ions act in a highly directional way.

A common problem with today's chips is that they generate heat when working. Some highly complex chips are so densely packed with transistors that they generate as much heat as a small light bulb. If tomorrow's superchips were built with today's conventional NMOS (N-type metal-oxide semiconductor) technology, they would get too hot to function properly.

The answer, it is said, lies in CMOS technology. The "C" stands for "complementary," indicating that N-type (for negative) and P-type (for positive) transistors are combined into a single element, whereas the dominant NMOS devices are built up from single N-type transistors. The result is that a CMOS chip uses only 10 percent of the power of an NMOS chip when operating and a mere 1 percent when standing by. The overall consequence is far less heat. Besides generating less heat, CMOS chips are better able to cope with variations in temperature, exposure to radiation and the extreme conditions that can be found in some heavy industry.

For many years, US firms ignored CMOS technology in favour of NMOS, because CMOS chips were considered too expensive, too slow and too big. It took the Japanese to demonstrate the advantages of CMOS, when they flooded the market with battery-operated products like watches and calculators which contained low-power CMOS chips. CMOS is also widely used in telecommunications, where chips can be powered by tiny voltages passed along telephone lines. By the late 1980s, it is predicted that CMOS circuits will account for about one-fifth of all chips – and the Japanese will turn out over half of them. Some electronics companies are working on so-called BIMOS circuits which combine the two technologies of bipolar and CMOS in a single chip.

CMOS is being widely used by a growing number of firms making made-to-order "custom chips." These come in two forms: "gate arrays" or "uncommitted logic arrays" (ULAs), and "silicon compilers" or "standard cells."

Gate arrays and silicon compilers are coming into their own because of the need for chip customers to work much more closely with chip manufacturers in the design of their increasingly sophisticated chip-using products. At the same time, chip manufacturers are facing rapidly rising costs in the design of entirely new chips: today, it is not uncommon for a team of 10–20 design engineers to spend two to three whole years on a new chip which might contain up to half a million devices. But with chips containing 8, 10 or 20 million devices on the way, it is clear that development costs will be phenomenal – and there may not even be enough chip designers in the world to do the necessary work.

Gate arrays or ULAs comprise a standard chip designed and manufactured with a set, or array, of electronic components such as logic gates already present but unconnected. When the customer has decided what the chip is to do, the appropriate connections are made to complete the circuit. The silicon compiling or standard cell technique takes the gate array approach a step further by keeping whole sections of already designed and tested circuitry in an electronic library. The new chip is then designed by computer utilizing these tried-and-tested sections arranged in the best possible way. The great advantages of gate arrays and silicon compilers are speed, cost and appropriateness for today's market, which is demanding more task-specific chips. Gate arrays are a fast-growing segment of the semiconductor industry, and sales are expected to increase four-fold by the end of the decade.

Another crop of companies (like Daisy Systems, Mentor Graphics and Valid Logic Systems) are developing the fast-growing technologies of computer-aided design (CAD) and computer-aided engineering (CAE). CAD-CAE promises to revolutionize the art of chip-making by enabling the non-expert to design chips incorporating silicon compilers and gate arrays. Chip-makers are also trying out robots, self-guiding vehicles and elaborate computer controls in their latest ultra-clean, automated chip-making facilities. The Japanese are ahead in automating chip production.

The search for ever-faster and more efficient chips has resulted in silicon's supremacy as the basic material for chips being threatened by gallium arsenide. This substance is formed by combining the scarce metal gallium with arsenic, the favourite Victorian poison. Gallium arsenide chips achieve faster operating speeds than silicon chips (up to ten times faster), and, rather like CMOS chips, they consume much less power, are more radiation-resistant and can operate in a wider range of temperatures. Gallium arsenide chip sales are expected to soar from

$30 million in 1985 to at least $1 billion a year in the early 1990s.

In fact, the performance of gallium arsenide chips appears to be so good at present that firms making equipment for the military are rapidly switching to the new material. Missile guidance, radar and electronic warfare are likely military uses, while gallium arsenide chips are also extremely suitable for telecommunications applications, such as cellular radio, microwave transmitters and fiber optics. There are still many technical obstacles that must be overcome before gallium arsenide is used as widely as silicon for standard VLSI chips, but there are indications that gallium arsenide chips could follow "Moore's Law," with a doubling of complexity every year into the 1990s and reductions in cost. Progress with gallium arsenide recently led IBM to abandon years of development work on Josephson junctions (tiny switches made from superconductors such as niobium), which were supposed to take over from silicon as the basis of chips.

But skeptics point out that many claims made for gallium arsenide in the past have not yet been realized. Gallium arsenide chips are still difficult and ridiculously expensive to make. And gallium arsenide semiconductor research – in which Japanese companies have a big lead – is still some years behind that of silicon. What's more, gallium arsenide is itself in theory under threat from so-called "superlattices" – semiconductors in which extremely thin layers of different semiconductor materials are combined into a single multilayer chip. Experimental superlattices in computer company and university labs have allegedly recorded speeds four times faster than those for gallium arsenide chips.

Gallium arsenide has also been used in experiments with optoelectronics, an emerging technology marrying optical processing and electronics which some say will produce the super high-speed machines of the future. An optical computer (of which prototypes already exist) would use light waves manipulated by lenses, mirrors and other devices to do the work of electrons and transistors in conventional computers. Meanwhile, optical disk storage technology, in which a laser is used to record information on a compact metallic disk, is undergoing great advances, especially in Europe, where massive amounts of data are now being put on a disk 30 cm in diameter. In February 1986, Bell Labs in the US announced they had established an optical computer research department.

Finally, mention should be made of so-called "biochips," about which there has been some speculation in the media. In fact, there are broadly two very different conceptions of what a "biochip" is, or could be. First, ordinary chips, protected by an unreactive inner layer and coated with an outer layer which responds to chemicals, could be used as sensing devices to operate in a biological environment. These might be more properly called "biosensors" and could, for example, monitor the condition of the human heart or other vital organs. Already scientists have developed bomb sniffers and drug detectors which operate in this way.

Second, it has been suggested that the electronic switches in silicon chips could be replaced by organic or genetically engineered protein molecules to create a full-blown "molecular computer" or computer-on-a-biochip. In "digital biochips," molecules would imitate the binary on/off functions of transistors, while in "analog biochips" enzymes might make possible a more graduated response to signals using their 3-D geometry. Such biochips – in theory, at least – would make possible even greater reductions in size and/or cost, increases in complexity and entirely new styles of data processing, more suited to high-level tasks.

Molecular or chemical computers might overcome the reliability and overheating problems that will occur when gallium arsenide and electron beam lithography are developed to *their* fullest extent. But beyond theoretical speculation, little practical work has been done to make the "biochip" a reality – although some research projects are now getting underway in the USA.

Supercomputers

Despite the amazing advances in microelectronics, your average personal computer may perform only 500,000 arithmetic operations per second. At the other end of the scale are *supercomputers*, which can perform 400 million or so operations per second (a million floating point operations per second is called a Megaflop). These giant number-crunchers cost between $5 million and $20 million to build, and only about 150 have been installed worldwide, mostly in government agencies and universities.

Although the operating speeds of supercomputers like the Cray X-MP and Control Data's Cyber 205 have tended to double on average every two years, they are still actually too slow for many potential tasks, such as complex computer simulations in the auto and aircraft industries which enable designers to see how their planned product will behave in certain conditions. But a new concept in computer architecture ("architecture" is the logical organization of a computer), namely parallel processing, is transforming supercomputers and bringing about improvements in performance and price reductions, which in turn are releasing pent-up demand for the new machines.

The problem with conventional computers using the dominant "von Neumann architecture" (named after the Hungarian–American mathematician John von Neumann) is that tasks have to be broken down into a sequence of steps and then laboriously executed one at a time. This means that the pace of computing is set by electronic circuit speeds, which in turn are limited by the agility of transistor switches, the speed of electricity and the excessive heat generated by hard-working chips. Parallel processing, on the other hand (as we saw with the Intel 432 and the Inmos "transputer"), is faster and cheaper because it enables the computer to execute the many different steps simultaneously.

US manufacturers Cray Research Inc. and Control Data Corporation are already selling supercomputers based partly on parallel processing. They and others, like Denelcor Inc. and ETA Systems, are also planning the next generation of supercomputers which could achieve 1000 million operations per second (that's 1000 Megaflops, or a Gigaflop). US sales of supercomputers are currently increasing by about 30 percent per year. The Japanese, meanwhile, launched their National Superspeed Computer Project in 1982, a $200 million effort which aims to produce a 10 Gigaflop supercomputer by 1989. This machine will incorporate parallel processing and massive memory capability. Recently Hitachi, one of the six major Japanese computer firms in the project, delivered a 630 Megaflop computer to Tokyo University.

Potential demand for such machines is enormous, as computer simulation becomes commonplace in scientific research, engineering, manufacturing, chemical processing, seismology, meteorology, bio-medical supplies and the semiconductor industry itself. Complex problems of fluid and aerodynamics can now be analyzed by computer rather than by laborious calculations and costly re-creations. Supercomputers are bringing about what Cornell physicist Kenneth Wilson called "the computerization of science." More and more experiments in physics, fusion research, biology and chemistry can now be simulated rather than actually performed.

In February 1985, the US-government-backed National Science Foundation announced it would spend an initial $200 million on supercomputers for US universities. Four campuses were designated National Advanced Scientific Computing Centers: the University of California at San Diego; the University of Illinois at Urbana-Champaign; Cornell University at Ithaca, New York; and Princeton University, New Jersey. Another 30 campuses will be connected to a high-speed computing network. Cray would supply the California and Illinois machines and ETA Systems the Princeton machine, while (to some people's surprise) IBM and Floating Point Systems of Oregon would supply the Cornell center.

NASA announced in April 1985 that it was building a supercomputer center for designing shuttle aircraft at Mountain View in Silicon Valley, to be opened in early 1987. This would do away with the need to build a series of prototypes for testing in wind tunnels because the supercomputers will be able to simulate aero-dynamic flows around space vehicles and their engines. The first Cray-2 (launched in June 1985) to be used in the NASA system is being run at speeds of 250 million operations per second (although it is capable of 1.2 billion), but by the 1990s it is expected that most supercomputers will have reached 10 billion operations per second (or 10 Gigaflops).

While the Japanese companies Fujitsu, Hitachi and NEC announced new supercomputer products in 1985, a threat to the long-term hegemony of supercomputers emerged with the arrival of the so-called "superminis," incorporating multiprocessors like Digital Equipment Corporation's VAX machines and the Convex Computer Corporation's C-1. These machines are slower but much cheaper than true supercomputers. Some have chips based on RISC (Reduced instruction set computing) architecture, which means simpler, faster circuitry. In addition, so-called "parallel processors" – smaller computers which work on different parts of a problem at the same time – have hit the market in the shape of products from Intel Scientific, Denelcor, Floating Point Systems, ELXSI and Thinking Machines.

In perhaps the most exciting applications area of all – computer graphics – supercomputers and high-performance processors are revolutionizing the art of animation. Strikingly lifelike pictures of natural phenomena such as landscapes, surfaces and motions can now be created by computer and remarkable feats of visualization performed. This is transforming film-making and, for example, flight simulation, and will have a major impact on science and medical teaching and industrial training. At the same time, graphics chips are becoming more sophisticated as the demand for graphics software packages grows.

The Fifth-Generation Project

The race to build supercomputers is only one aspect of the even bigger race to build "Fifth-Generation" computers. The general idea is that the first generation of computers was based on valves, the second on transistors and the third on integrated circuits; the fourth and current

generation is based on VLSI chips, while the fifth will pull together all the state-of-the-art technologies in chip-making, memories, CAD, parallel processing, software, vision systems and speech recognition in order to create the "intelligent" talking, listening, thinking machines of the future. As *Business Week* recently remarked, "The next generation of advanced supercomputers will make today's machines look like hand-held calculators."

In early 1981 came news that the Japanese had launched something called the Fifth-Generation Computer Project. The aims of this ten-year, $500 million research effort were spelt out in a 28-page report from the Japan Information Processing Development Center (JIPDEC) and again at a conference held in Tokyo in October 1981. Japan planned to produce by 1991 the first of a new generation of what it calls "knowledge information processing systems" (KIPS), which would rapidly supercede conventional computers. In particular, they would be capable of automatic language translation (a priority for the Japanese), document preparation by voice-activated typewriters, expert professional advice, and decision-taking on the basis of logical inference.

The project is organized by way of a consortium of eight major Japanese firms, who were brought together in April 1982 to form the

Institute for New Generation Computer Technology (ICOT), which is located on the twenty-first floor of an office block in Tokyo. The dynamic director of ICOT is Kazuhiro Fuchi, who picked the 40-strong team of young computer whizz-kids upon whose cooperative efforts so much depends. These young researchers share ideas freely, but return to the eight firms weekly to keep them abreast of current research.

The actual technology goals of the Fifth-Generation Project are many and various, but basically they depend upon a series of "scheduled breakthroughs" in computer hardware and software, the development of expert systems, and natural language understanding by machines; this whole area is now referred to by the Japanese as "knowledge engineering" or "knowledge information-processing" (hence KIPS). Fifth-Generation computers will have to support very large knowledge bases, work at very high speeds, perform complicated logical inference operations as fast as current computers perform arithmetic operations, and develop a machine–user interface that allows significant use of natural speech and images.

All expert systems consist of three main parts: a subsystem, which manages the knowledge base needed for problem-solving and understanding; a logical inference subsystem, which decides what knowledge is useful and relevant for solving the problem at hand; and methods of interaction between human and machine that are "natural" and comfortable for the user.

Under the Fifth-Generation plan, knowledge will be stored electronically in a large file known as a relational database. This will provide greater flexibility, and it is hoped to expand it to handle tens of thousands of logical inference rules and 100 million objects – in other words, it will be capable of storing the entire *Encyclopaedia Britannica*.

Second, it is hoped that Fifth-Generation computers will be able to perform 100 million or perhaps even 1000 million logical inferences per second (LIPS), compared with only 100,000 LIPS in today's computers. Already, an individual workstation built by the Japanese using the simpler language PROLOG is capable of 1 million LIPS.

Third, the aim is to develop a human–machine subsystem which incorporates language understanding and image processing. This is a most difficult area of artificial intelligence research (see next section), but the Japanese researchers aim to create a machine translation system with a vocabulary of 100,000 words and 90 percent accuracy, combined with an image processing system capable of storing 100,000 images. In each of the three subsystems, hardware and software development will go hand-in-hand.

The Fifth-Generation Project is proceeding in three planned phases.

The first three years were spent learning the state of the art, and forming the concepts and building the hardware and software needed for the later work. The fruits of this research – including a computer capable of making 200,000 LIPS – were demonstrated in Tokyo in May 1985. The second phase of four years is being devoted to engineering experiments, building prototypes and seeing how the subsystems operate together. The third and final phase of three years will concentrate on advanced engineering, building the final prototypes, further systems integration work and the development of commercial products.

In their somewhat evangelical text, *The Fifth Generation: Artificial Intelligence and Japan's Computer Challenge to the World* (Addison-Wesley, Reading, Mass., 1983), Edward Feigenbaum and Pamela McCorduck argue that the Fifth-Generation Project amounts to a technological Pearl Harbour and the United States had better wake up before it is too late. Knowledge is the new wealth of nations, they say, and if the Japanese succeed in taking over world leadership in knowledge processing systems, they could dominate the world in every other way. Yet the Japanese plan is also inspired, they say, by a "noble vision of the future" – a future where knowledge is accessible to anyone, anywhere, any time, a democratic future that works.

Skeptics say the Fifth-Generation Project is risky and over-ambitious. It will be hard to maintain momentum and meet the prescribed goals on time. Japanese weaknesses in software will hold back progress. On the other hand, even if only 10 percent of the project's goals are achieved, it will probably have been worth it in terms of technological spin-off and new commercial products.

The US response has been fairly swift. In 1982, 12 major US corporations came together to form the Microelectronics and Computer Technology Corporation (MCC), a non-profit joint venture, to pool resources for long-range research (see chapter 1). Many top firms including IBM, Intel, Texas Instruments and Cray Research stayed out of MCC, but the mere fact that 12 firms are cooperating is a bold new departure from the American way. MCC, which is based in Austin, Texas, has a budget of $75 million a year and a staff of 250. Its first research projects are on semiconductor packaging, CAD/CAM, computer architecture, software and AI.

In the same year, 13 US chip firms formed another non-profit research consortium, the Semiconductor Research Corporation (SRC). The SRC is based at Research Triangle Park in North Carolina, from where it commissions research in universities, spending around $30 million a year. In addition, the Pentagon's Defense Advanced Research Projects Agency (DARPA) and its Strategic Computing Plan (SCP) are pouring millions of dollars into research on advanced supercomputers and artificial intelligence. The object is to develop technologies that will make possible a new breed of "smart" weapons, intelligent missile-guidance systems and military robots such as drone aircraft and unmanned submarines and land vehicles. The Pentagon's Strategic Defense Initiative ("Star Wars") project is giving a further massive boost to supercomputer research.

In Europe, the EEC has responded to the fifth-generation challenge by establishing the ESPRIT (European Strategic Programme of Research in Information Technology) research project, while the British government in 1983 launched the five-year $500 million Alvey Program of research in advanced information technology. Through 1984 and 1985, a whole series of Alvey contracts were awarded in the four main areas covered: VLSI chips, software engineering, expert systems and the man-machine interface. Amid growing concern that Alvey was not really working, the UK government in March 1986 set up a high-powered committee under Sir Austin Bide to advise on the program's future.

Toward Artificial Intelligence (AI)

Academics have been studying and developing "artificial intelligence" ever since a meeting which gave birth to the term at Dartmouth College in 1956. But in the main they have little in the way of commercial products to show for their efforts. There is as yet nothing to rival the human brain, which was 10 *trillion* circuits packed into an area the size of a small cabbage.

Early computers were actually referred to as "electronic brains," but if nearly 30 years of AI research has achieved anything, it has put to rest this naive analogy. Rather, the analogy has broadened and matured as scientists have learned more about the nature of human intelligence and thought. There is very little agreement, in fact, as to whether computers can or should be made to "think," but one thing researchers are agreed on is that the recent advances in microelectronics have given a tremendous boost to AI work and the development of useful AI systems.

The progress with supercomputers and the Fifth-Generation Project itself are obviously highly relevant to artificial intelligence, but specifically AI research covers the broad areas of expert systems, robotics (see chapter 6), speech recognition, image processing and the attempt to discover the rules by which humans think (sometimes called "applied epistemology"). AI research in the USA is associated primarily with four universities: Carnegie-Mellon, Massachusetts Institute of

Technology, Stanford and Yale. The most interesting and best supported research area has been that to do with expert systems, and recent successes – such as the discovery of an elusive molybdenum deposit in Mt Tolman, Washington, with the help of an expert system called "Prospector" – have created a wave of new interest in AI and a rush to form new AI companies.

To develop an expert system, AI researchers basically spend years picking the brains of experts, trying to extract their knowledge and to understand the way it is organized, so that it may be reproduced. What differentiates the operation of an expert system from that of a conventional system is its flexibility. A standard computer program specifies precisely how each task must be performed. An expert system tells the computer what to do without specifying how to do it, by using special languages such as LISP or PROLOG that allow a computer to manipulate symbols rather than numbers. The "how" and the "why" of the system are kept separate, allowing new data to be added at any time without altering the program.

In an expert system, the computer therefore exercises judgment, and it does so by using certain intellectual rules of thumb, a process known as "heuristics." Heuristics has also been described as the "art of good guessing." It enables experts, human or machine, to recognize promising approaches to problems and to make educated guesses when necessary. The first expert system based on heuristics was Dendral, devised in the late 1960s by Edward Feigenbaum and Joshua Lederberg at Stanford University as a method of analyzing the chemical structure of compounds. The latest version of Dendral is Genoa, a system that is in regular use with organic chemists today.

A better known expert system is Mycin, developed by Edward Shortliffe at Stanford in the mid-1970s. An interactive system that diagnoses bacterial infections and recommends therapy, Mycin is an expert system that uses "backwards chaining" – in this case, to see if the evidence supports the diagnosis. An offshoot of Mycin called Puff helps diagnose lung problems and is in regular use in San Francisco; another system called HELP is in use in hospitals in Salt Lake City and Elmira, New York; while Stanford has set up a national computer network called SUMEX-AIM to help AI work in medicine. Perhaps the most advanced medical system is Caduceus, developed by Dr John Myers and Harry Pople at the University of Pittsburgh. Caduceus can diagnose unrelated diseases, using heuristics to narrow down the field and drawing on information in its large memory bank. However, it began field-testing at four hospitals only in November 1986.

Some commercial systems up and running include Xcon, developed by Digital Equipment Corporation to help its sales force establish the exact needs of each customer; Ace, a system devised by Bell Labs to help trace and analyze faults in telephone cables; and Prospector, developed by SRI International for the US Geological Survey, which contains the knowledge of many expert prospectors and is partly credited with the Mt Tolman find. General Electric has a system for detecting faults in diesel locomotives, while Teknowledge Inc. of Palo Alto and the French oil company Elf Aquitaine both have in use expert systems which advise oil rig operators on what to do when the drill bit gets stuck.

Boeing has an expert system to help engineers design aircraft more quickly, while GE has a system to diagnose jet engine faults. Honeywell's Mentor system helps diagnose problems in air conditioning, and Composition Systems Inc. of Elmsford, New York, has a system that supervises the layout of newspapers. Applied Expert Systems (APEX) of Cambridge, Massachusetts, is marketing the Plan Power System for sorting out personal pension plans, while the Campbell Soup Company lays claim to a soup production expert system which embodies the knowledge of retiring soup production engineer, Aldo Cimino, accumulated over 44 years of experience with Campbell's soup cauldrons.

The UK too has seen an upsurge of interest in expert systems. Examples on offer include systems for car fleet operators, travel agents and farmers, to help them detect fungal wheat diseases. Expert system "shells" or "tool kits" – skeleton programs on which customers can hang their own experts' knowledge – such as the Xi system from Expertech of Slough, Berkshire, have achieved substantial sales.

In the early 1980s, a number of large corporations like IBM, Xerox, Schlumberger and Hewlett–Packard announced substantial investments in AI research. The Pentagon's DARPA, SCP and SDI initiatives were soon pouring millions more into this esoteric area. By the mid-1980s, corporations were falling over themselves to buy up or buy stakes in the 200 or so small US companies working on AI-related products, especially expert systems in the new field of "knowledge engineering."

Despite the slump in semiconductors, there was also a mad rush to form start-up companies specializing in systems using LISP, the AI computer language, for such areas as financial services, manufacturing, defense and AI "tool kits." IntelliCorp of Mountain View, California, had sold more than 500 copies of its $6000 KEE tool kit by February 1986, while the other members of the AI "Gang of Four" – Teknowledge Inc., Inference Corp and the Carnegie Group Inc. – had each sold between 50 and 150 copies of their tool kit programs. But in many fields progress in transferring expertise from expert to program was often very slow.

In a different but relevant sphere, work on speech recognition and "natural language processing" is also making slower-than-expected progress. The human voice, with its accents and inflections, is presenting a tough challenge to speech technologists everywhere. Some speech recognition systems are available, but they have vocabularies of hundreds, rather than thousands, of words, and it will be many years before vocabularies are built up to a point where natural recognition is possible.

Machine translation (MT) is showing more progress in the short term, but speech recognition has even greater commercial potential in the long term. Voice recognition systems could make electronic mail, messaging and computer communications much more user-friendly. Natural language processing systems which handle instructions in English would be a boon to all computer users, while voice-activated typewriters (VATs) would transform office work. As computing power and memory become cheaper, building versatile voice recognition and natural language processing systems will become more cost-effective.

Many argue that AI has been dreadfully "oversold" and that it may once again fail to deliver the goods. Despite the ballyhoo, not many usable expert systems have been produced (less than 40, compared with over 200 experimental projects), and those that are in use are closer to old-fashioned sequential computer programs than anything resembling real "intelligence." There is growing skepticism about whether expert systems will ever become truly "expert." Current technology seems best suited only to diagnosis or classification problems whose solutions depend primarily on the possession of a large amount of specialized knowledge. If AI is to even approximate human intelligence, scientists will have to teach computers not to take things literally, to "learn" and to use "common sense."

But can computers really be taught to "think" like human beings? AI's leading apostles, like Herbert Simon of Carnegie-Mellon and Edward Feigenbaum of Stanford, say that it is perfectly possible, while Douglas R. Hofstadter of the University of Indiana argues in his book, *Gödel, Escher, Bach* (Basic Books, New York, 1979), that "intelligent" machines are just not on. Joe Weizenbaum, Professor of Computer Science at MIT, has, through a stream of articles and books like *Computer Power and Human Reason* (W. H. Freeman, San Francisco, 1976), kept up a barrage of criticism of the AI fraternity, accusing them not only of absurd over-optimism in their research aims, but also of paving the way to totalitarianism. AI enthusiasts, he says, confuse information with knowledge, knowledge with wisdom and thinking with calculating.

More recently, Hubert and Stuart Dreyfus, in their book *Mind Over Machine* (Free Press, New York, 1986), have argued that information-processing technology will probably *never* resemble the human brain, which works by intuition rather than rule-following and data processing, which are strictly low-grade mental operations. Machine intelligence will probably never replace human intelligence simply because we ourselves are not "thinking machines." AI, the Dreyfuses write, has a long history of inflated claims: "AI has failed to live up to its promise, and there is no evidence that it ever will."

In reply, Marvin Minsky, also a professor at MIT, has argued that, just as the theory of evolution has changed our view of life, so artificial intelligence will change our view of the mind. As we learn more about mental processes, we will develop new perspectives on "thinking," "feeling" and "understanding" which do not see them as magical faculties. This may lead to the building of new machines and, in turn, to new ideas about what constitutes human intelligence. It is therefore wrong to claim that computers will *never* resemble human brains because, in truth, we do not know enough about either. But advances in VLSI, parallel processing and the whole computer revolution will take us many giant steps nearer finding the answers.

3 Silicon Valley: Home of High-Tech Man

Origins – Secrets of Success – Semiconductor Industry Trends – The Japanese Chip Challenge – Problems in Paradise – Here Comes High-Tech Man

At the south end of San Francisco Bay, some 50 miles from downtown San Francisco, lie the fertile flatlands of Santa Clara County, sandwiched between two ranges of low hills. The area was known as recently as 1950 as the "prune capital of America." Today the fruit orchards are mostly gone, replaced by a gridiron of futuristic factory units, new freeways and rows of new apartment blocks which march toward the shimmering horizon. This is "Silicon Valley," home of the world's microelectronics industry and the "high-tech" culture that may provide a model for the society of the future.

Silicon Valley, as it was first dubbed by local newsheet editor Don C. Hoefler in 1971, has been responsible for many of the major innovations in electronics in recent years, from the hand-held calculator, digital watches and video games to home computers and the microchip itself. Some 1100 of the firms in the American Electronics Association – over 60 percent of the total – are located in California, and over 600 of them are in the Valley. Massachusetts, with only 112 firms, is placed second. Sales of the US electronics industry overall will reach $400 billion by the end of the 1980s, making it America's fourth largest industry – and Silicon Valley will be responsible for the lion's share.

Since the prune farmers moved out, Silicon Valley has rapidly developed into the ninth largest manufacturing center in the USA. Over 300,000 people work in the heart of the Valley. Despite recent troubles, new jobs are still being created at the rate of 20,000 a year. San José, at the southern end of the valley, is America's fastest-growing city – it jumped from twenty-ninth to eighteenth place in the big city league between 1970 and 1980. In their book, *Silicon Valley Fever* (Basic Books, New York, 1984), Everett M. Rogers and Judith K. Larsen cite reports that there are over 6000 PhDs in Silicon Valley and over 15,000 resident millionaires, making it the greatest concentration of scientific brainpower and new wealth in the USA, if not in the entire world.

Origins

Most people trace the origins of the Silicon Valley semiconductor industry to the decision of William Shockley, co-inventor of the transistor, to move west in 1955. In that year, Shockley returned to his home town of Palo Alto (so named by Gaspar de Portolá after a scraggly pine tree, or "tall stick") at the north end of the valley to found Shockley Semiconductor Laboratories, the first such company in the San Francisco area. Prior to that, the infant semiconductor industry had been centered around Boston and Long Island, New York. But in fact, the seeds of Silicon Valley had been sown much earlier – by the founders, and key figures in the development of, Stanford University.

Stanford University, opened in 1891, was created in memory of the son of Senator and Mrs Leland Stanford, who donated $20 million and 8800 acres at Palo Alto for the project (see figure 3.1). Senator Stanford was the Central Pacific railroad magnate who drove the golden spike into the track near Ogden, Utah, which first linked the East with the West. Technically oriented from the start, it wasn't very long before Stanford began to have an effect on nearby industries: in 1912, for example, Lee de Forest of the Federal Telegraph Company, Palo Alto, invented the vacuum tube amplifier, his research partly funded by Stanford University.

A brilliant 1920 graduate, Frederick Terman, was to play a vital role in Stanford's – and thus Silicon Valley's – development. Forced by illness in 1924 to forgo an MIT professorship and remain in California, Terman became Stanford's professor of "radio engineering." In the early 1930s two of his students were William R. Hewlett and David Packard, later to become the founders of Palo Alto's largest and perhaps most famous electronics firm, Hewlett–Packard. Terman encouraged the duo, even to the extent of enticing them back from the East in 1938 with the help of fellowships, accommodation and commercial work.

By 1942 Hewlett–Packard employed 100 people, although it was 1950 before the firm added the next hundred. But rapid expansion (on Stanford University land) followed as the company moved from audio and electronic instruments into computers and microchips. By 1983 Hewlett–Packard manufactured 5000 products, had 68,000 employees and was ranked 110th in the *Fortune* 500. Dave Packard is now worth over $2 billion and Bill Hewlett about $1 billion, making them the two richest entrepreneurs in America today (Rogers and Larsen, *op. cit.*).

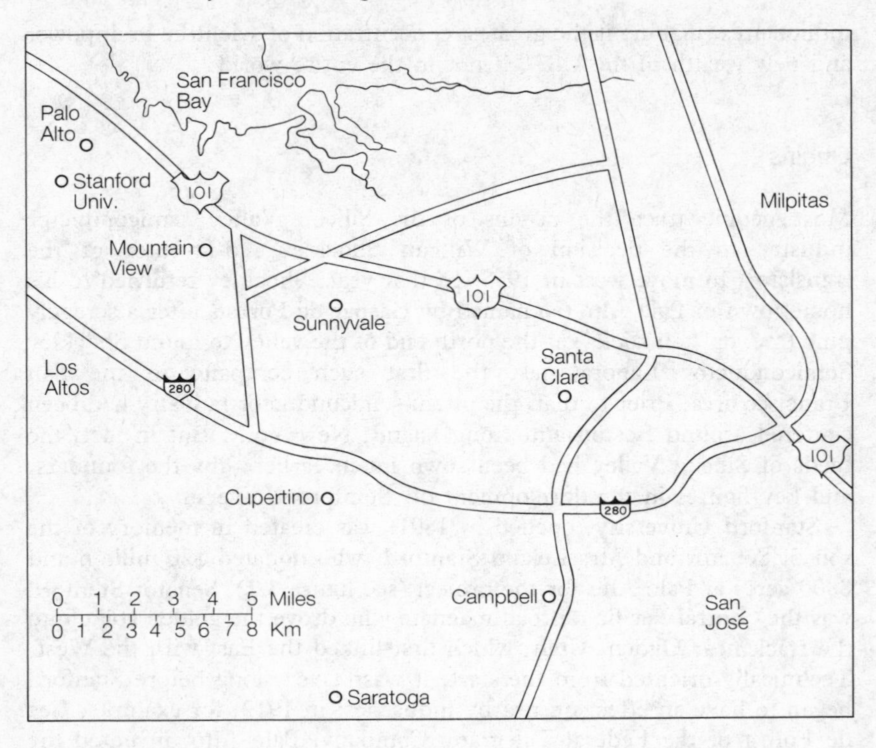

Figure 3.1 Silicon Valley.

Another Terman brainwave was the Stanford Research Park, created in 1951. This was the first such high-tech industrial park located near a university, an idea that has since been widely copied in the USA and Europe. By 1955 seven companies had moved into the park; by 1960 there were 32, and today there are 90 companies leasing all 655 acres. Income from the leases has produced a handy income for Stanford over the years and has enabled it to become one of the world's greatest research universities.

Shockley Semiconductor Laboratories, the first actual semiconductor company in the Valley, was of course the brainchild of William Shockley. Shockley represented the new breed of postwar scientific entrepreneurs. Astute and determined to make a million, it was he more than anyone else who was responsible for originating the entrepreneurial tradition of Silicon Valley. Unfortunately for him, most of the brilliant young electronics engineers he recruited from the East had similar ideas:

in 1957 eight of them, including Robert Noyce and Gordon Moore (later to found Intel), left to form Fairchild Semiconductor. People were shocked at the time by this seeming "betrayal" by the "Shockley Eight," but it became the pattern for future company development in the Valley.

Fairchild and the "Fairchildren" (people who worked for Fairchild in the early days) have had an even greater influence in the Valley in succeeding decades, as Michael S. Malone argues in his book, *The Big Score: The Billion Dollar Story of Silicon Valley* (Doubleday, New York, 1985). It is said that, at a conference of key figures in the semiconductor industry held at Sunnyvale in 1969, less than two dozen of the 400 people present had *not* worked at some time for Fairchild. By the early 1970s, it was possible to trace some 41 companies founded by former Fairchild employees. "There can be no doubt," write Ernest Braun and Stuart Macdonald in their book, *Revolution in Miniature* (Cambridge University Press, 2nd edn, 1982), "that the location of Fairchild explains much of the present concentration of the [semiconductor] industry in California." But, they say, it is only part of the story. The San Francisco Bay area had other attractions besides Fairchild. . . .

Secrets of Success

The sunny San Francisco Bay area of northern California has a climate that many consider to be ideal. Proximity to the Pacific ensures that average temperatures remain within the 60–80°F range all the year round. Humidity is rarely high and damp almost unknown. When the inland temperature does rise on a hot afternoon, San Francisco's famous fog rolls in through the Golden Gate Bridge to cool things down. This refreshing daily occurrence has been described as "nature's air conditioning."

San Francisco itself is one of the world's most beautiful cities. Quite unlike most of urban America, it attracts superlatives from all who visit it and has figured in numerous songs, novels and films. The wider Bay area also seems to be everybody's favourite residential location. Parts of it are getting over-developed now and house prices are sky-high, but northern Californians still enjoy a higher standard of living than most other places in the world plus an excellent quality of life.

It's therefore not surprising that the area has continued to attract young, well-educated Americans looking for the good life, as well as Hispanic migrants looking for jobs. This has created a pool of highly skilled scientific and technical personnel – a vital requirement for a new, expanding industry like microelectronics – and a reservoir of unskilled labour eager for assembly-line work. With local universities cranking out science and engineering graduates who wished to stay in sunny California, entrepreneurs just setting up in Silicon Valley and existing firms relocating there in the 1950s and 1960s had a ready supply of locally available brainpower. Without it, Silicon Valley might never have got started, say Rogers and Larson.

Likewise, military funding played an important role in the early days of Silicon Valley. At one time in the 1950s and early 1960s, Department of Defense purchases accounted for as much as 40 percent of total chip production. Firms like Transitron, for example, were wholly dependent on military purchases. Today the percentage is more like 7–8 percent. The change occurred in the late 1960s and 1970s as the computer and consumer markets really took off. Also, the military's requirements came to diverge from those of commercial chip users, and the industry became less willing to make special arrangements. There is a continuing debate about the precise role of military purchases in the origins and growth of the semiconductor industry, with skeptics pointing out that much military research money was wasted on blind-alley, ill-fated projects.

In addition to the meteorological climate, the Californian business climate has proved especially favorable to the growth of the Silicon Valley semiconductor industry – an industry in which most of the important innovations have come from small firms, either in the form of spin-offs from established firms (like Intel, who gave us the first microprocessor) or brand-new start-ups (like Apple, who gave us the first personal computer). Components of this business climate include a good communications infrastructure; ready supplies of venture capital; an abundance of skilled, mobile and mostly non-union labor; and a high value placed on the individual, entrepreneurship and money-making. Semiconductor industry characteristics include a preponderance of small firms; high spending on R&D; a premium on being first with a new product; high rates of job mobility; networks of contacts between firms; and intense competition to attract and hold industry "stars." It is easy to see how industry characteristics and the local business climate have interacted to form a potent mix, to create a "synergy" in Silicon Valley which has proved unique and unstoppable.

Many of the key individuals and companies in Silicon Valley came from the eastern USA or Europe. They came to escape the restrictions of hidebound cultures and companies, attracted by the opportunity and openness of Californian society. They were also the *type* of people who would be so attracted and repelled. As Valley venture capitalist, Don Valentine, put it, "The East is large companies and rigid structures. The individual doesn't fit well in them. California is the frontier: unstructured economically, socially and institutionally, and above all with a real commitment to personal net worth. That's why Silicon Valley grew up here and not somewhere else." Or as one Intel manager told me during a visit in 1978, "It was something to do with the nine-to-five attitude and not feeling part of a team in the big corporations. In a small-team set-up, you don't have to refer things to some goddam committee all the time. You just get right on with the job yourself. We get English guys coming out here all the time saying they can't take the bureaucracy."

Being able to attract a certain type of person was therefore one secret of Silicon Valley's success. As Braun and Macdonald point out, "The [semiconductor] industry is one of people rather than companies in that the prosperity of these companies is, perhaps more than in any other industry, dependent on the abilities of individual experts." Moreover, from the beginning, semiconductor expertise has been regarded as the property of individuals rather than companies. Skills required by companies were acquired by hiring the key individuals. Company progress would largely depend upon creating the conditions in which these "stars" would come up with ideas for new products.

With the emphasis on individuals rather than companies, most electronics engineers have had no qualms about "job-hopping" from firm to firm, taking with them their "turn-key" knowledge. It is often said in Silicon Valley that if you want to change jobs you simply take a different exit off Highway 101 on the way to work in the morning. Many employees have done just this, with the result that job-hopping is rife in the Valley: annual turnover rates of 30–50 percent in firms are not uncommon, although they have been falling recently. Head-hunting agencies exacerbate the problem for established employers by enticing key workers away from firm A with offers of salary increases and better conditions at firm B.

Each such job change provides an opportunity for silicon stars and superstars to advance their careers, to gain further knowledge and (perhaps) to be part of a winning team. This process results in a high degree of cross-fertilization in the Valley which, although it is disliked by many employers, has been vital to the Valley's success. But the primary motivation for job-hopping is money, with 15 percent being the standard raise upon moving. Shortages of skilled personnel and high housing costs in the Valley mean that qualified individuals can rapidly increase their income by frequent moves. There have been cases in the past of certain valued individuals being offered huge sums as "transfer fees," along with salary increases, bonuses, interest-free loans, shares and stock options.

A high degree of labor mobility may be good for new or expanding firms and the Valley as a whole, but it's bad news for many established employers who have to go to great lengths to hold onto the employees they've got. High pay, help with housing costs, sabbaticals, company gyms, swimming pools, saunas and hot tubs, free drinks and frequent parties are all part of the corporate culture created by firms like Rolm, who have themselves changed from labor poacher to gamekeeper in the space of a few years.

Profit-sharing schemes and stock options are now widespread in Valley companies. Laudable though this may be, cynics see them simply as another device to entrap valued employees, and stock options (often relinquished on leaving the company) are sometimes referred to as the "golden handcuffs." Having said this, there are some old-established firms, like Bell Laboratories and Xerox, who have traditionally taken a relaxed view of the loss of personnel, and others, like IBM (in the Valley at Santa Teresa), whose corporate culture is so strong that they rarely lose anybody.

The microelectronics industry was founded on individual expertise, and it advances through the exchange of expertise and information. Job mobility is one way of bringing about transfers of knowledge. Networks of contacts – often built up through job changes – are another. Networks are important in the Valley because each firm wants to stay up with, or get one jump ahead of, the competition. In an industry based on rapid innovation, success is highly dependent on the exchange of information about how to solve problems, what techniques have already been tried and so on.

The close proximity of firms in the Valley makes for the rapid transfer of information about developments in other firms through informal networks of contacts. Information might be traded over the telephone between friends and ex-colleagues or exchanged more directly at one of the Valley's favourite watering holes, such as the Wagon Wheel Bar, Mountain View (once described as the "Fountainhead" of the Valley), the Cow Girl Bar at the Sunnyvale Hilton or the Peppermill just off Highway 101. Attempts have been made by some employers to restrict the divulging of information by employees and ex-employees, but they have usually failed. Firms do sometimes formalize arrangements through licensing the use of patents, but of far greater importance is the exchange of information via the informal networks. Without them, the vitality of the Valley would suffer.

Silicon Valley is justifiably famous for entrepreneurship, particularly in the form of "spin-off" firms set up by defecting employees. The trend was started by the "Shockley Eight," and since then hundreds of talented engineers and marketing experts have taken their skills and ideas and set up on their own. Sometimes they have felt genuinely frustrated in their old firm, but more often their main aim has been to make a fast buck before they burn out. Most firms in the Valley are spin-offs of one kind or another, or spin-offs of spin-offs, but some started from scratch. The classic story is Apple Computer, which was started by the young Steve Jobs and Steve Wozniak in a garage in 1976, achieved sales of $200 million four years later and had grown into a $1 billion corporation just eight years later. Nolan Bushnell built up Atari even quicker, although it was later to founder (under new ownership) as the market for video games declined.

But while media coverage of Silicon Valley has tended to focus on these and other spectacular success stories, little has been heard of the many hundreds of firms that failed to make the grade. A survey by Professor Albert Bruno of the University of Santa Clara found that a relatively high 95 percent of Silicon Valley start-ups survive the first four years, but 25 percent fail to survive their "adolescent transition" and collapse in the second four years. Of 250 firms founded in the 1960s, 31 percent survived, 32 percent had been acquired or merged and 37 percent had gone bust.

Nowadays, the cost of setting up a new semiconductor facility is so enormous that budding entrepreneurs must work hand-in-hand with financiers or so-called "venture capitalists." Risk capital was always more plentiful on the West Coast – indeed, that's another reason why Silicon Valley got started where it did in the first place. But the semiconductor industry, with its unique requirements, has been largely responsible for creating a new breed of person, expert in both electronics and finance, whose job it is to assess the potential of proposed companies and product ideas. The San Francisco Bay area is now famous for its firms which specialize in this kind of work.

As John W. Wilson explains in his book, *The New Venturers: Inside the High-Stakes World of Venture Capital* (Addison-Wesley, Reading, Mass., 1985), venture capitalists – or "vulture" capitalists, as they are sometimes unkindly called – gamble by putting money into new or growing high-technology firms with the potential for rapid growth. They do this on the assumption that some will go bust and some will survive, but it's the spectacular Apple-style "winners" they're really after, because these can repay the initial investment many times over. Venture capitalists not only act as gatekeepers, deciding which companies will get started; they also participate in the day-to-day management of the company in order to help safeguard their investment. A good example is Ben Rosen, chairman of the Sevin Rosen group, who was behind two of the computer industry's most sizzling start-ups: Lotus, producer of the best-selling 1–2–3 spreadsheet program, and Compaq, the third largest personal computer maker after IBM and Apple.

Most venture capitalists aim to make an average annual return of 35 percent. But some of the more successful individuals and funds have notched up annual returns of 70 percent or more. This, and the reduction by Congress of long-term capital gains tax from 49 to 28 percent as a result of semiconductor industry pressure in 1978, produced a flood of venture capital in the early 1980s. There was for a while concern in the Valley that it had become *too* easy to start a new company and that a high proportion of start-ups were failing. But from about mid-1984 on, rates of return went down and venture capitalism seemed less attractive as it became harder to find promising young high-tech companies in which to invest.

With so much emphasis on electronics whizz-kids and venture capitalists, commentators have often overlooked the role of marketing and public relations in the growth of successful Valley companies. After the initial technical breakthroughs in the 1950s, 1960s and early 1970s, commercial factors became just as important, if not more important, than the technology itself. Why should anyone buy firm A's widget rather than firm B's? Who was to know whether one was better than the other? The customers had to be told. . . .

Into the void stepped the PR men, epsecially a philosophy graduate by the name of Regis McKenna. McKenna had a knack of crafting compelling images for companies, and he played a key role in creating the reputations of National Semiconductor, Intel and Apple – indeed, he is credited with inventing the Apple name and logo and the whole idea of a "home personal computer." Another little job he tackled was the launch of Intel's microprocessor, the first computer-on-a-chip.

In his book, *The Regis Touch* (Addison-Wesley, Reading, Mass., 1985), McKenna describes how he works at giving companies a "personality" and talks of "positioning" companies in the market. He sees PR as integral to overall business strategy, which he regards as a process rather than an event. L. J. Sevin, whose Mostek company (first absorbed by United Technologies and later sold to the French group, Thomson) was once on a par technically with Intel, says he was "clobbered" by McKenna's campaign to position Intel as the market leader. "I didn't realize PR mattered so much [in this industry] until it was too late," he told *Fortune* magazine.

Semiconductor Industry Trends

In recent years, the Silicon Valley semiconductor industry has seen output growth averaging 20 percent a year. But progress has usually been in fits and starts, with booms being followed by slumps. Also, the industry is not without its problems, such as widespread price-cutting, rising investment costs, the rapid obsolescence of much production technology and the shortening of product lives. As a result, there have been some famous bankrupticies, numerous takeovers or mergers, and a number of moves toward collaboration and partnership which would have seemed highly unlikely a decade ago.

The demand for microchips – memories, microprocessors and custom-made circuits – is largely dependent on the demand for chip-based and chip-using products and is therefore greatly affected by changes in the US economy. Thus, the 1974–5 semiconductor industry recession ended with the late 1970s boom, which in turn came to an end with the slump in semiconductor demand in 1980–2, when sales actually declined by about 10 percent overall. The rapid growth in demand for personal computers and VCRs in particular created boom conditions in Silicon Valley in 1983: orders leapt, prices of standard chips rose after a steady decline, and widespread shortages of key devices were reported. The only weak spot was in demand for ROM chips, owing to the decline in popularity of video games.

With more chips being incorporated into mainframe computers, telecommunications products, automobiles, industrial control equipment and military hardware, the boom continued right through 1984, making it the best year of the industry's history: worldwide semiconductor shipments were up an incredible 46 percent to about $27 billion, and some Valley firms such as Advanced Micro Devices and Signetics notched up sales growth of 70–80 percent. As firms rushed to increase production capacity, the capital spending boom which began in 1983 became a bonanza for production equipment makers: during 1984, the top ten US semiconductor firms increased capital spending over 1983 by 67 percent. One after another, established chip firms reported record profits, and it was also a very good year for new start-ups, especially the more specialist "niche" companies.

But behind the euphoria, there were fears that the industry was simply into another phase of its familiar "boom–bust" cycle, based on a failure to balance supply and demand. The cycle works like this. Worried about reported shortages and long order books, chip users place extra orders with different suppliers. The suppliers simultaneously lay down extra capacity in response. But because of long "lead times" (maybe two years), when the new production lines finally come on stream, there are too many chips on the market chasing too few buyers. The order book, or the "book-to-bill" ratio – a measure of orders received over orders shipped – declines and new investment is postponed. As Intel's Gordon Moore once put it in a widely quoted statement, "The balance of supply and demand in this industry lasts for about 35 minutes, between the end of one boom and the beginning of the next recession."

And so it was that, toward the end of 1984, the familiar warning signs of cancelled orders and a decline in the industry's "book-to-bill ratio" appeared. From all-time highs of 1.65 in December 1983 and 1.53 in January 1984, the book-to-bill ratio plunged to 0.67 in November 1984, meaning that orders in hand amounted to only two-thirds of recent sales (see figure 3.2). As one analyst put it, "At the end of 1984, the semiconductor industry fell over a cliff." It was the same story in Europe, where the book-to-bill ratio dipped below 1.0 in the first quarter of 1984, falling even further to 0.65 in January 1985.

In the first few weeks of 1985, there was a distinct change of mood in the Valley as the full implications of the downturn in the personal computer market sank in. Some leading semiconductor companies had started to defer capital projects and stopped hiring new workers, but many were taken completely by surprise. No one had foreseen the depth of the slump that was to come in 1985. Rampant over-capacity produced a sudden glut of chips which sent prices tumbling. In February, Valley companies announced the first round of job cuts – Intel dismissed 900

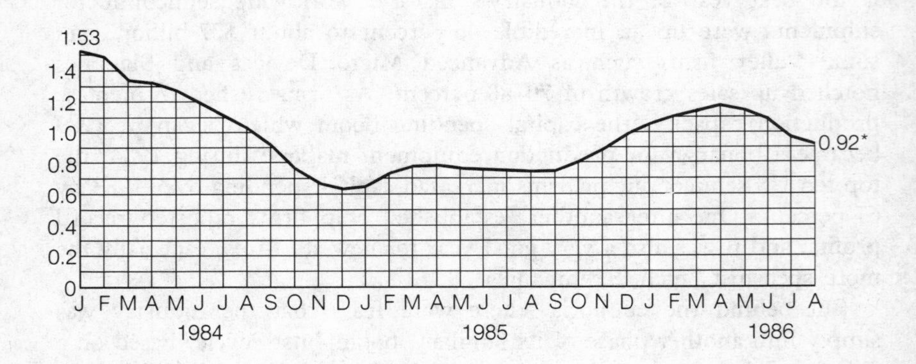

Figure 3.2 The book-to-bill ratio.
(Source: Semiconductor Industry Association)

workers and put two plants on a four-day week. Texas Instruments, Zilog, Signetics, Monolithic, Mostek and many others followed suit, laying off workers and/or announcing temporary plant shut-downs.

By March it was clear that Silicon Valley was in for a very rough time indeed. Falling chip demand was exacerbated by a strong dollar, aggressive price-cutting by the Japanese and competition from US and foreign-owned plants in the Far East in places like Taiwan, Korea, Singapore and Malaysia. It turned out that even IBM's PC was not really American at all: $625 worth of its $860 manufacturing cost was accounted for by components made overseas. Final import–export figures for 1984 showed a US deficit on electronics trade of $6.8 billion, and the continuing surge in imports of components and peripherals pointed to a really massive trade deficit for 1985.

Chip prices tumbled 40–50 percent, and by April industry spokesmen were talking about a "bloodbath" as the accountants' red ink flowed. Standard 64K RAMs were selling for $0.75, down from $3.50 a year earlier (they finally bottomed out at $0.30). The price of 256K RAMs was down from $12 in early 1984 to $5 in mid-1985. While the Japanese continued to add capacity, the surge of cheap Japanese imports led to growing calls for protection and pressure for action against unfair trading practices (see next section).

By June, with the book-to-bill ratio stuck at around 0.75, analysts were talking about the semiconductor industry "slump" which seemed to go beyond the normal boom-and-bust cycle. They suggested deeper reasons for the crisis in computers: over-indulgence on the part of companies who needed time to digest the technology they had recently bought; technical bugs; lack of compatibility; technophobia on the part of executives; and a general disenchantment with all information technology.

In October, the first ray of hope appeared in the form of a small rise in the book-to-bill ratio to 0.82. The Semiconductor Industries Association (SIA) came out with some bullish predictions for 1986 and the inventories of chip-buyers came down, but skeptics pointed out that over-capacity was still a chronic problem. The overall balance sheet for the semiconductor industry in 1985 went like this: 64,000 workers laid off (nearly 20 percent of the workforce); worldwide sales down 17 percent, US sales down a massive 31 percent.

The new year – 1986 – brought better news. In January the book-to-bill ratio rose above unity to 1.04 for the first time in 17 months. In February it hit 1.10, and in April it reached a 12-month high of 1.18. Signs of a significant upturn emerged in the form of workers being recalled and prices rising. But it was a false dawn: the book-to-bill ratio fell through May to 1.09 in June, 0.96 in July and a mere 0.92 in August. The longest depression in the industry's history showed every sign of persisting.

Rising investment costs and over-expansion are the bane of the semiconductor industry. As production technology becomes more automated and more sophisticated, costs rise steeply. For example, a state-of-the-art chip production line might cost $100–200 million, anything from four to eight times what it cost only a few years ago. Some devices for special production processes have gone up 20 times in cost. Yet firms are tempted to invest heavily in the latest equipment because the "experience curve" dictates that production costs drop roughly by 30 percent for every doubling of volume. If prices stay firm, there is an even greater incentive to increase capacity. US chipmakers added 90 chip production lines in 1983 and another 200 in 1983, increasing the total capacity of the industry by 30 percent.

Firms have coped with the problem of the boom–bust cycle and rising investment costs in different ways. Some have diversified out of semiconductors into products like laser-scanners, automated teller machines and specialist medical equipment. Others have concentrated on proprietary designs or have gone in for "nichemanship" – finding small niches in the market which provide regular sales and above-average profits. Application-specific custom and semi-custom chips also provide some protection from violent swings in demand for standard or commodity chips, although by mid-1986 it was still not clear whether even the top semiconductor companies would be able to handle the transition away from the standard chips to custom and semi-custom chips.

But the major response – and perhaps the industry's best hope for the future – is evident in the increasing number of tie-ups, mergers and alliances which are occurring between suppliers and end-users. This trend toward "vertical" or "backward" integration actually started some years ago, with Schlumberger's purchase of Fairchild (generally considered to have been a disaster) and the absorption of Mostek by United Technologies back in 1979. Gould also took over American Microsystems Inc. (AMI) in 1981, the year that Intel signed a collaboration agreement with Advanced Micro Devices (AMD). More recently, the trend has become a scramble as firms have rushed to form partnerships and to broaden their range of products. As Stephen T. McClellan, author of *The Coming Computer Industry Shakeout* (John Wiley, New York, 1984), puts it, "A powerful convergence of forces – technological, economic, societal – is recasting the industry from top to bottom."

Thus, in recent years IBM has purchased 20 percent of Intel, a leading chip-maker (although it sold some on) and 100 percent of Rolm, which makes specialist telecommunications equipment. Wang has bought into VLSI Technology, NCR into Ztel, Sperry into Trilogy, Magnetic Peripherals into Centronics and Control Data into Source Telecomputing. Western Union has bought into Vitalink, while AT&T has gone overseas to purchase 25 percent of Olivetti, the expanding Italian office equipment maker. RCA has teamed up with Toshiba – together, they ranked eleventh worldwide in chip sales in 1985. Later in the same year, Toshiba reached a wide-ranging agreement to cooperate in semiconductors with Siemens of West Germany. With technology moving fast, product lives shortening and the cost of developing new products increasing, not even giants like IBM can afford to develop and produce a full range of products.

The convergence of electronics, computing and telecommunications means that firms who once thought of themselves as just being in semiconductors or mainframe computers or software or telecommunications now find they are all in the same information-processing industry. They are expected to offer a complete range of information-processing equipment – this is known as "one-stop shopping" – and teaming up allows them to do this. Thus Burroughs, even before its merger with Sperry in 1986, already supplied small computers, medium computers, software, telecommunications equipment and peripherals, all of which had been made by someone else. This increased collaboration reflects concern about IBM's entry into every sector of the market and more especially fear of Japanese competition in a fast-growing industry which is expected to be worth an amazing $1000 billion by 1990.

With the semiconductor industry firmly incorporated into the much larger information-processing industry, analysts predict that we will end up with a few giant-sized, vertically integrated suppliers like IBM and AT&T, who can provide everything users require; a much larger number of second-tier firms, who can supply integrated computer systems, parts of which are made by different manufacturers; and a large number of small, specialist firms developing new hardware and software products. Most of these will still be based in Silicon Valley.

Finally, there has been other evidence of a shakeout in the industry as competition has increased. In recent years, the Valley has seen some spectacular failures, mostly connected with the personal computer boom. Osborne Computer, for example, had all the ingredients of success and one of the first portable personal computers on the market in 1981. Huge sales were recorded in 1982, but by September 1983 founder Adam Osborne had to file for protection under Chapter 11 of the US

Bankruptcy Code. In the same year Victor Technologies also crashed, Atari was in deep trouble and Texas Instruments had to pull out of the personal computer market altogether. Even Apple had a number of bites taken out of it. In 1984 it was the turn of Gavilan Computer and Convergent Technologies, both producers of portable personal computers, to make the headlines with spectacular marketplace failures. Eagle Computer, TeleVideo, VisiCorp, Trilogy and Diasonics all ran into trouble. As one Valley wit put it at the time, "The personal computer industry has reached a new chapter in its history: Chapter 11."

Apple, the symbol of Silicon Valley success, was shaken to the core in March 1985 when company president John Sculley (hired from Pepsico in 1983) announced the shut-down of its factories for a week. This followed a fall in Apple's share of the personal computer market from 22 percent in 1982 to 11 percent in 1984. As Apple's earnings growth stopped, its share price declined, executives left ("Apple Turnover" according to one headline), the ad budget was slashed and new products delayed. Co-founder Stephen G. Wozniak left to form a new company called CL9 (as in "Cloud Nine").

In June 1985 Apple closed two out of its three USA factories, cutting 1200 from its workforce of 5500. At the same time, Apple co-founder and main driving force Steve Jobs gave up responsibility for day-to-day operations, marking the end of an era at the company. Losses mounted, and in September 1985 Jobs resigned as chairman and left altogether, taking five managers with him. Amid a great deal of public mud-slinging between Jobs and Sculley – which led to a law suit, eventually settled out of court in 1986 – Jobs announced plans to form a new company called Next Inc. to make sophisticated microcomputers for the university market.

Summing up his career, *Time* magazine described Jobs as "the brash, brilliant and sometimes bumptious brat of Silicon Valley, a symbol of its high-tech genius and fabulous sudden wealth. . . . Jobs was the prototype of a new American hero – the irreverent and charismatic entrepreneur."

The Japanese Chip Challenge

Fear of the Japanese began to grip the Valley in 1978 when the first Japanese chips arrived in significant numbers. US producers could not keep up with demand during the late 1970s boom because they had reduced investment during the 1974–5 recession. They were caught off guard, so much so that by the end of 1979 the Japanese had captured

42 percent of the market for 16K RAMs, the latest generation of memory chip at the time.

Then came a bombshell. Richard W. Anderson, a general manager at Hewlett–Packard, made a presentation to a Washington conference in March 1980 in which he reported Hewlett–Packard's findings on the comparative quality of US and Japanese chips. During 1979 the firm had purchased 16K RAMs from three US and three Japanese producers: they found that the *best* of the American products had a failure rate *six times* that of the lowest-quality Japanese firm. Industry observers were amazed. Even though Anderson was to report a dramatic narrowing of the quality gap two years later, his initial presentation sent shock waves through the Valley. US firms suddenly realized they had a fight on their hands.

Events in 1981 were to turn fear of the Japanese into a phobia. The 64K RAM, successor to the 16K RAM, was widely seen as the new workhorse of the computer industry, the standard memory chip of the future. Potential sales were enormous. A 64K RAM contains 65,536 memory cells, each of which can be reached independently, rather like a telephone network. Being a RAM (random access memory) rather than a ROM (ready-only memory), the new chip could be erased like a miniature blackboard and re-used. The 64K RAM was destined to become the basic building block of computers, telecommunications equipment and a vast array of electronic products – and by the end of 1981 the Japanese were producing no less than 70 percent of them. Influential industry figures started to panic. Even the Governor of California, Jerry Brown, was moved to warn that Silicon Valley "could become the next Detroit."

The Valley had certainly been surprised by the suddenness of the Japanese assault. Intel had produced the first 1K RAM only in 1970, and throughout the 1970s US firms like Intel, Mostek, Texas Instruments and Motorola had continued to lead the world in memory devices, through the 1K, 4K and 16K generations. But, while they were struggling to execute complicated new designs for packing 64K of memory into a tiny chip, the Japanese chose a simpler and more conservative approach, which basically meant doing a bigger version of the 16K RAM. Its simplicity meant fewer problems, cheaper production and greater reliability. In one fell swoop, the Japanese had "leapfrogged" the US industry, and their policy of rapid price-cutting (prices for 64K RAMs fell from $28 to $6 in one year) enabled them quickly to grab a huge slice of a key market.

Price-cutting deprived US producers of the early profits they had expected to make on their first 64K RAMs and wrecked their long-term plans. And by gaining early experience with 64K RAMs, the Japanese were in a far better position to tackle the 256K RAM market. Their domination of the memory market allows them to set industry technical standards not just for memory chips, but for other chips as well. As Jerry Sanders, chairman of Advanced Micro Devices (AMD), put it, "Memories are where we learn the technology for all the other semiconductor devices. If you're not in memory, you're at a big disadvantage."

Around 15 US producers were in memories up to the 16K level. Only five stayed the course with 64K RAMs, although IBM and Western Electric produce large numbers of them for their own use. As a result of their strength in memories, there are now five Japanese companies in the world top ten chip producers: Hitachi, NEC, Fujitsu, Toshiba and Matsushita. The normally self-effacing Japanese are so confident of continued domination in memories that they are predicting that they will completely control the market for 256K RAMs (which is expected to be worth $14 billion) by 1990, and probably that for 1 megabit, 4 megabit and 16 megabit RAMs as well.

As with all Japanese miracles, the Japanese success in semiconductors was planned right from the start. Faced with growing environmental pollution and increasing energy costs, the Japanese decided to "target" information technology and the knowledge industry as early as 1970. In 1971 the Japanese government published its futuristic "Plan for an Information Society" and MITI, the giant Japanese Ministry of International Trade and Industry, was instructed to ensure that Japan became a world leader in computers within the decade. The key to success in computers would be success in semiconductor chips, which MITI correctly predicted would become "the rice of industry."

In retrospect, the Japanese were in a strong position to take advantage of advances in microelectronics, particularly the VLSI revolution. Leading Japanese computer firms like Hitachi, Fujitsu and Nippon Electric Co. (NEC) belong to *kieratsu*, or large families of companies which make a vast range of products. Firms within the *kieratsu* thus provide a ready-made market for semiconductor devices. The *kieratsu* also have ready access to relatively cheap, long-term finance and are not bothered by stockholders or backers looking for quick returns. This is crucial in the case of VLSI, which requires very large sums of money for initial capital investment. As one observer put it, "The Japanese can hold their breath for a long, long time."

In addition, the all-important 64K RAM introduced the need for planning and team effort, for which a more stable workforce was needed. The high mobility of labor in Silicon Valley – normally hailed as a great advantage – began to look like a big drawback. The Japanese had the continuity and the experience of teamwork. MITI in the mid-1970s also sponsored massive research and development work on VLSI technology, the results of which were to be shared with major Japanese computer firms; and the Japanese PTT, the Nippon Telephone & Telegraph (NTT), also shared its research with major suppliers, who mostly happened to be the same firms. This is hardly the American way of doing things. As George Scalise, vice-president of AMD, explained at a congressional hearing into US–Japan competition in July 1981, there are "major structural differences" between US and Japanese industry, and the Japanese approach constitutes "the greatest threat to the long-term viability of the US semiconductor industry."

Furthermore, the Japanese have proved themselves to be second to none in perfecting manufacturing techniques to give maximum quality and reliability. The making of semiconductors calls for meticulous attention to detail and cleanliness, at which the Japanese excel. VLSI production is also highly automated, with increasing use of automatic test equipment, which the Japanese have been particularly good at developing, and robotics, in which the Japanese have a world lead. In fact, the Japanese in the early and mid-1980s were ahead in automating their semiconductor industry, spending proportionately more on, say, automatic test equipment. In 1984 US manufacturers of all chip-making equipment were losing market share – for example, Japanese firms had grabbed about a third of the vital market for photo-aligners and VLSI logic testers from virtually nothing in 1979. Moreover, Japanese machines have often proved to be more reliable and easier to set up. Both these factors increase yield and thus directly affect prices and profitability.

Japanese semiconductor firms also went on a capital spending spree. In 1984–5, they were ploughing back some 25 percent of their sales revenue into new production facilities, compared with about 14 percent in US firms. But this represented a three-fold increase in spending in barely more than three years. Overwhelming confidence on the part of the Japanese that they could subsequently cut prices and increase market share has led to frequent accusations in the USA that the Japanese were "dumping" chips on the US market. As early as July 1982, the Justice Department launched an investigation, and in 1983 the Semiconductor Industry Association published a report by Robert Galvin, chairman of Motorola, which not only alleged unfair trading practices on the part of the Japanese, but also charged that the whole Japanese set-up of cartels, shared R&D costs, government subsidies and predatory pricing "would be clearly illegal actions in this country."

As US firms began to fight back, there was new evidence that the Japanese were no longer ahead on quality. In April 1984, for example, major chip-user Sperry announced that only three firms had made the grade into their "100 PPM Club" for suppliers of chips with less than 100 defective parts per million: the US firms Intel, National Semiconductor and Motorola. The US share of the world memory market in 1984 actually rose to about 35–40 percent, and there were grounds for optimism on the 256K RAM front. Even in the 64K RAM race, Texas Instruments and Mostek climbed back into the top five (to join Hitachi, NEC and Fujitsu).

With growing sophistication in the chip market – and customers demanding special kinds of memory rather than just more memory – US firms benefited from being nearer in geography and language to the big customers. This enabled them to develop more specialized products and to find lucrative "niches" in the market. Moreover, although the Japanese were strong in semiconductors, they had yet to break into the much bigger market for integrated computer systems. In 1984 Japan still had only around 10 percent of the world market for data processing equipment, while the USA had 80 percent. The Japanese had no one to match IBM, who seemed determined to dominate.

But by early 1985, continuing Japanese success in semiconductors in the midst of the most serious slump in Silicon Valley's history led to renewed pressure on the US government. Final figures for 1984 showed that the US share of the world market for chips was falling, while Japan's was rapidly rising. From a chip trade surplus with Japan in 1980, the USA had run up a trade deficit of $800 million in 1984. Despite calls by Japan's prime minister, Yasuhiro Nakasone, for trade liberalization, it was pretty obvious that Tokyo had no intention of opening up its home market to US companies.

The fear that the Japanese would dominate the emerging market for megabit memories, following their successes with 64K and 256K RAMs – which gave them 70 percent of the total D-RAM market – led the Semiconductor Industry Association (SIA), in June 1985, to file a complaint under section 301 of the 1974 Trade Act with the International Trade Commission (ITC) and the Commerce Department. There were also clear signs that Japanese chip companies (especially NEC) were planning to mount a major challenge in microprocessors, a market in which Japan already had a 30 percent share.

In response to a complaint from memory maker Micron Technology of Boise, Idaho, the ITC ruled in July 1985 that Japan *was* dumping 64K RAMs – and this was confirmed by the Commerce Department in April 1986. In September 1985, Intel, Advanced Micro Devices (AMD) and

National Semiconductor filed anti-dumping petitions over the pricing of Japanese EPROMs – and the complaint was upheld in November 1985. In the same month, the newly formed Electronic Industries Association of Japan (EIAJ) hit back at US allegations in a submission to the ITC. But in January 1986 the ITC ruled against the Japanese once again, charging this time that they had dumped 256K RAMs.

Friction grew between Washington and Tokyo, but in March 1986 talks between US and Japanese state officials reached no agreement. With the Japanese by then controlling 90 percent of the memory market, the Commerce Department ruled in April 1986 that the Japanese were illegally dumping memory chips and announced the imposition of import duties. Under growing pressure, the Japanese finally signed the US–Japan chip accord in July 1986, by which they agreed to raise the prices of devices made in Japan. This upset US chip users and led Japanese firms to switch production to the US and Europe. Perhaps most significant of all, Fujitsu announced it planned to acquire 80 percent of Fairchild and merge its US operations. The US government launched an investigation.

Despite its obvious deleterious consequences, it can still be argued that the Japanese assault on chip-making has had overall a salutary effect on Silicon Valley. Prices have been kept lower, the pace of innovation has increased, and above all, US semiconductor firms have been forced to improve their quality. US firms have also learned a lot from Japanese manufacturing and management techniques. Industry leaders have been forced to reconsider their attitude toward collaboration and government assistance. Thus by December 1986 – when new figures confirmed the Japanese as chip leaders with 38 percent of world sales *vs.* 36 percent for the US – the US government had before it two proposals to help the US semiconductor industry regain its supremacy: the SIA's $1 billion "Sematech" research project in advanced chips and the Defense Science Board's $1.6 billion aid package for the US semiconductor industry. A year or two earlier, suggestions like this would have been non-starters.

Problems in Paradise

Silicon Valley faces other problems besides Japanese competition. Environmental pressures have resulted in serious air and water pollution – including at least one major disaster – while there is growing concern over the occupational health and safety record of this supposedly "clean" industry. Crime has soared, and the Valley has been rocked in recent years by a series of "spy" scandals involving the Soviets and the Japanese. Since 1982 the industry has also been confronted with the problem of Pentagon-inspired high-tech export curbs, which have reduced lucrative sales to certain foreign countries.

Like most things that work in America, the Silicon Valley microelectronics industry was not planned: it just happened. But the same free-wheeling business climate that helped create the phenomenon in the first place could now be sowing the seeds of its own destruction.

Shortage of Space

Silicon Valley has a fundamental problem of space. Bounded by ranges of hills, San Francisco Bay and the Pacific Ocean, there is only a limited amount of land in the Valley and a severe shortage of land available for development. This has pushed up land prices and thus the costs of new industrial developments and much-needed housing. When average house prices hit $100,000 in the early 1980s, shortages of key personnel were widely reported and many Valley companies made plans to set up plants in cheaper areas, such as Texas, Colorado, Washington and Oregon.

Rogers and Larson say the unplanned way in which the Valley developed from north to south has allowed the more attractive North County communities of Palo Alto, Mountain View and Sunnyvale to cream-off the top microchip companies, with their profits and their large, landscaped sites; South County communities around San José, like Gilroy, were left to pick up what companies they could and to provide low-cost housing for the Valley's manual workers. Housing the poor is bad news for an area's property tax base: the costs of providing services like the police, fire protection, education and welfare are high relative to income. As a result, affluent white North County has grown even richer at the expense of South County, where lower-class whites and almost all the non-whites are housed. Because of high housing costs in the north, manual workers face a long commute every day on the Valley's over-burdened freeways.

In the absence of a public transportation system – over 90 percent of Valley residents go to work by car – there are fears that car-derived air pollution in the form of photochemical smogs will become commonplace. The number of cars in the Valley continues to grow at an explosive rate; the average car commute had grown to 15 miles and 46 minutes by 1984 and was expected to increase rapidly. In fact, there are real fears that the main north–south freeways will soon seize up altogether on occasions. No one thought to plan a mass transit bus or rail network for Silicon

Valley because nobody knew it was going to develop in the way it did. And while the unbridled pursuit of self-interest by individual entrepreneurs and municipal authorities made things happen, in retrospect, it also stored up problems for the future (Rogers and Larson, *Silicon Valley Fever*).

Pollution and Occupational Health Hazards

Car exhaust gases are one cause of air pollution; the large quantities of chemicals, toxic materials and gases used in chip production are another. You will find no belching smokestacks reminiscent of "Rust Bowl" industries in Silicon Valley, but every day about 9 tons of "reactive organic gases" are released into the atmosphere by Valley semiconductor companies, according to the Bay Area Air Quality Management Agency. In order to comply with the federal Clean Air Act, this figure will have to be reduced substantially – perhaps to 3 tons a day. But the companies say they cannot meet the costs.

By far the most serious pollution problem in the Valley involves the contamination of groundwater by leaking underground chemical storage tanks. In December 1981, workers excavating near the Fairchild chip plant at South San José discovered chemicals in the soil. A subsequent check revealed that no less than 58,000 gallons of solvent had leaked out of a flimsy fiberglass underground tank, undetected. Some 13,000 of the missing gallons were the toxic solvent, trichloroethane (TCA). Less than 2000 ft away was Well 13, an aquifer belonging to the Great Oaks Water Company, which had been supplying contaminated water to thousands of local residents in the Los Paseos neighborhood. Samples had been taken from Well 13 in the summer of 1980, but tragically they had been lost before testing. About 400 plaintiffs – nearly half of them children – began proceedings against Fairchild, alleging the company had caused miscarriages, the premature deaths of babies, birth defects, cancer, skin disorders and blood diseases. The plant was closed down, and Fairchild spent $12 million on monitoring wells and soil replacement.

After the Fairchild leak, local officials throughout the Valley hastily drew up new ordinances to cover the underground storage of chemicals. Numerous leaks came to light elsewhere. In April 1982, the California Regional Water Quality Control Board initiated an underground tank inspection program. Of the first 80 sites visited, 64 (80 percent) showed evidence of soil and/or water contamination by solvents, acids, metals, resins and fuels leaking from tanks, sumps and pipes; 57 of these were classified as serious and in need of further investigation. The Environmental Protection Agency (EPA) has also embarked upon a $500,000 project to study air and water pollution in Silicon Valley by toxic chemicals, but even this does not cover potential hazards such as the accidental rupture of cylinders containing toxic gases like arsine, phosphine and diborane: their release into the atmosphere could result in poisoning on a mass scale, and no community in the Valley is equipped to handle such a disaster.

In January 1985, Dr Kenneth W. Kizer of California's Department of Health released the findings of his two-year study into the impact of the Fairchild leak on the Los Paseos area. He found that miscarriages were running at twice the normal rate and there were three times as many birth defects – 30 percent of all pregnancies in Los Paseos ended either in miscarriages or birth defects. In response to the Kizer report, California's Governor George Deukmejian announced an urgent inquiry into water pollution by toxic chemicals used in the semiconductor industry. A lively debate ensued between a local environmental group, the newly formed Silicon Valley Toxics Coalition, and company-sponsored organizations.

Recent research by Dr Joseph LaDou, of the University of California's School of Medicine in San Francisco, has revealed the extent to which the "clean" image of the microelectronics industry is misleading: in fact, it poses far greater health and safety problems for its workers than had been previously thought. Writing in *Technology Review* (May/June 1984), LaDou showed that company health and safety records indicate an unusually high incidence of occupational illnesses in the semiconductor industry – 1.3 illnesses per 100 workers, compared with 0.4 in general manufacturing industry in 1980. Similarly, 18.6 percent of lost work time in semiconductor companies in 1980–2 was due to occupational illness, compared with only 6.0 percent in all manufacturing industries. Moreover, compensation statistics showed that 47 percent of all occupational illnesses among semiconductor workers in California resulted from exposure to toxic chemicals – *twice* the level of all manufacturing.

One undoubted reason for this state of affairs is the fact that this is a newish industry, making new products with new processes. As Ron Deutsch of Signetics points out, "There are chemicals in use we still don't know enough about and they may be more harmful than we know. We're still learning. Even consultants in the field aren't sure how these things react." This suggests that the situation will improve in time as companies learn more about possible hazards. Others argue the opposite, saying that, because many of the industry's workers are still young, it may be years before the true horrors come to light. Moreover, in an industry where small, entrepreneurial companies are prevalent, the incentive to cut corners is strong. As specialist Dr Dan Teitelbaum puts it, "There are a lot more problems in smaller plants. They are just not good with health and safety. It's a myth that this is a clean industry."

Security Risks

Whatever happens, microchip companies are likely to face increased costs not only in combating pollution and improving occupational health, but also in the form of higher security costs, as firms struggle to reduce theft, counterfeiting and espionage – now multi-million dollar industries themselves in California.

Chips are small and easily pocketed. Their origin can be disguised, and they can be copied. The result is that millions of stolen, counterfeit and even defective chips find their way onto the "grey market" every year and into the hands of competitors, unsuspecting customers and foreign governments. Since many crimes remain undetected or unreported, nobody knows the true extent of the problem. But it is a problem that can be deadly serious, as became clear in September 1984, when it was revealed that thousands of potentially defective Texas Instruments chips had actually found their way into advanced US nuclear weapons systems.

In June 1982, Silicon Valley was stunned to learn of the arrest in Santa Clara of six senior executives of the Japanese firms Hitachi and Mitsubishi, for conspiring to steal documents, parts and software from IBM. They and 14 others were subsequently accused of paying a total of $648,000 to a firm of consultants called Glenmar Associates for material and trade secrets they knew to be the property of IBM. It turned out that Glenmar Associates was an FBI front – the luckless Japanese with their "shopping list" of secrets had been entrapped in a "sting" operation that later became known as "Jap-scam." At about the same time, an employee of National Advanced Systems, a subsidiary of National Semiconductor, was arrested for receiving goods stolen from IBM, and in September 1982 IBM fired three of its high-ranking officials for trying to sell secrets.

When the Hitachi people came to trial on criminal charges in San Francisco in February 1983, they pleaded guilty. The company was fined the maximum $10,000 and two employees, $14,000. IBM's civil suit against Hitachi was finally settled out of court in October 1983. Hitachi agreed to pay IBM substantial damages and all of IBM's legal costs (estimated at several million dollars). IBM also won the right to inspect all of Hitachi's new products for five years to ensure that they are not using IBM secrets. In another settlement, in January 1984, National

Semiconductor agreed to pay IBM $3 million out of court after IBM documents were found at National's headquarters in Sunnyvale. Meanwhile, IBM filed suits against five more former employees for using proprietary technology.

"Jap-scam" and these related events shocked many in the Valley who were still unaware of the extent of Japanese industrial espionage. It also showed the lengths to which IBM was now prepared to go to guard its secrets, especially against the so-called PCMs (plug-compatible manufacturers, who make equipment that plugs into IBM equipment). But industrial espionage was hardly a new phenomenon in the microelectronics industry of northern California. After all, the creation of "spin-off" companies by former employees taking away secrets was what helped make the Valley great in the first place. One could even say that industrial espionage – in the form, for example, of hiring key employees away from a competitor – is part of the industry's way of life. Yet the borderline between legitimate business activity and illegality these days seems to have become increasingly blurred.

Various intelligence-gathering techniques are used, both by domestic and international competitors and by foreign governments. Many are perfectly legal, such as listening to "bar talk" in order to pick up scraps of information. Firms can also learn a lot about their competitors by gaining access to public records using the 1966 Freedom of Information Act. "Predatory" hiring – tempting employees away from a competitor with generous cash offers – is perfectly legal, although the disclosure of secrets could result in a civil suit alleging theft of "intellectual property," as we saw in the 1985 case of Texas Instruments vs Voice Control Systems. Indeed, legislation to protect intellectual property in the form of semiconductor chip designs – the Semiconductor Chip Protection Act – was finally approved by the House and the Senate in October 1984. It provides ten-year protection for the photomasks used in chip production.

Less savory but still legal is the familiar technique of the phoney employment ad. A firm advertises and interviews prospective employees from other firms about their work, maybe even getting them to produce samples. More often than not, no job offer materializes. Bogus companies can be set up to offer collaboration or to elicit information through joint discussions or – in the case of the Soviets in particular – simply to buy new equipment or products derived from military equipment. Further on up the scale into illegality are bribery, blackmail, sexual favors and – increasingly common – outright theft in the form of pilfering or planned break-ins to steal secrets.

The Japanese are not the only foreigners active in industrial espionage. The Soviets and other Eastern bloc countries maintain massive worldwide intelligence-gathering operations because they are desperate to keep pace with US technology. In his book, *High-Tech Espionage: How the KGB Smuggles Nato's Strategic Secrets to Moscow* (Sidgwick & Jackson, London, 1986), Jay Tuck estimates that the USSR has no less than 20,000 agents trying to obtain US high-technology secrets through such tried-and-tested techniques as dummy companies and bogus research institutes. Fully two-thirds of the major Soviet weapons systems are reckoned to be based on US technology. The KGB even has its own "wish list," a thick catalog of high-tech items – called, appropriately, *The Red Book* – which it would like to acquire.

California, with its high concentration of defense, aerospace and communications firms, is a prime target of foreign industrial espionage, and Silicon Valley, with its hundreds of leading-edge electronics firms – many working on classified military applications – has become the bulls-eye. Soviet espionage operations in the Valley are directed from the Soviet consulate on Green Street, San Francisco, a building that appears to be over-staffed, with about 70 personnel, and bristles with antennae, no doubt used to listen in on corporate conversations.

In 1982 the Reagan administration became concerned about the extent of the Soviet intelligence effort after a sophisticated Soviet marker buoy was picked up in the Pacific and found to contain chips from Texas Instruments. The previous May, a Soviet spy named Kessler was stopped boarding an aircraft at Los Angeles with two suitcases containing $200,000 worth of secret radar equipment taken from the Hughes Aircraft Corporation. (The celebrated Harper case was to follow in 1983). Reagan announced extra funds for the US Customs Service to mount Operation Exodus, designed to halt the flow of high-tech goods abroad, and US negotiators were instructed to take a tough line at the July 1984 talks of COCOM (Coordinating Committee for Multilateral Export Controls), the Paris-based committee of 15 Western countries which agrees embargoes of sensitive technologies that might reach the Eastern bloc.

In subsequent months, a fierce tug-of-war developed between the Department of Defense (backed by the President) and the Department of Commerce (largely echoing the views of exporters) over the proposed tightening-up of the Export Administration Act 1979, due for revision anyway by Congress and the Senate. Partly as a pre-emptive strike, and partly in order to demonstrate the bureaucratic problems involved in further restricting high-tech exports, the Department of Commerce proposed complicated new rules to narrow the scope and destination of sensitive US goods in January 1984. But the Defense Department's case had been immeasurably strengthened by the dramatic seizure, in Sweden in November 1983, of a sophisticated Digital Equipment Corporation VAX computer, en route to the Soviet Union. It had originally been sold to a phoney South African company, who had exported it to West Germany and thence to Sweden. Similar seizures took place in Vienna, Austria, in 1984 and in Luxemburg in 1985. New, tighter export control regulations involving distribution licenses came into force in July 1985.

Meanwhile, US manufacturers led by IBM began a campaign to warn the Reagan administration of the consequences for US business of tough export controls. They found active support among some European governments, still smarting over the 1982 Siberian gas pipeline dispute. Some British MPs, opposing the Thatcher government's Export of Goods (Control) Order of 1985 – which restricted the export from Britain of computer kit which might find its way to the USSR or its allies – charged that the USA was indulging in a new form of high-tech "imperialism." In trying to prevent the latest technology from falling into Eastern bloc hands, the US government was also ensuring that the industries of its European allies lagged behind.

As it became clear that Britain's own fifth-generation project, the Alvey Programme, could be severely jeopardized by embargoes on the export of software and advanced computer languages, the suspicion grew that the US government was in fact helping some major US corporations to increase their domination of the world information technology market.

There was also concern that the US administration, through COCOM, was restricting high-tech exports from Europe to China, while allowing US companies greater freedom to gain a foothold in this last great untapped market of the world. The wrangle ended in January 1986, when COCOM lifted most restrictions on exports from Europe to China. But it was replaced by a new argument over "extra territoriality" – the desire on the part of the US government to apply US export control laws and "audit" the operations of US-owned and/or US-supplied firms in Europe.

Here Comes High-Tech Man

Silicon Valley has developed a culture all its own. Some say that it provides us with a glimpse into the future, and that Silicon Valley is a kind of prototype for the evolving high-tech society. This may or may not be true, but certainly the combination of "chip capitalism" and California has produced a unique way of life that may provide clues to all our futures. Boosters of Silicon Valley marvel at its technological achievements, the new wealth created and the high-tech goodies

emerging from it. Critics point to the seamy side of things in the Valley – like South County – and the high social costs of technological progress, charging that high-tech culture represents the triumph of me-first individualism over the common good.

Silicon Valley people work hard – perhaps too hard. Engineers often put in 15-hour days and seven-day weeks. The competition between firms is so intense that survival may well depend on being first to market with a new product. Many are out to simply get rich quick. Only work matters – and nothing else. In their book, *Silicon Valley Fever*, authors Rogers and Larsen describe one electronics engineer who boasted of working for 59 consecutive days without a break. An executive at National Semiconductor with a rare blood disease carried on working with an intravenous bottle and pump at his side, even in the car. Charles Peddle, president of Victor Technologies, is quoted as saying: "I put my three vice-presidents in hospital within the past two years, and three of the top four officers in Victor lost their families through divorce in the past year."

At this inhuman pace, many engineers and executives "burn out" at an early age – rather like sports and pop stars. Indeed, many expect to do so, and this increases the pressure to "make a pile" within ten years. A recent MIT study by Professor Louis Smullin found that many computer and electronics whizz-kids were "washed-up" by the time they were 35 or 40 and were "fighting a losing battle to keep from falling behind intellectually." Each year, about 10,000 electronics and electrical engineers – or 5 percent of the US total – transfer out of the field altogether. In Silicon Valley, this is called "dropping off the edge." It goes on all the time, but is rarely talked about.

During the 1985 semiconductor slump, it is said that Silicon Valley psychiatrists were inundated with clients suffering from "silicon psychosis" – a stress syndrome affecting success-driven types living through the industry downturn. Many executives, who had known only growth and expansion, developed a negative self-image and experienced a loss of faith in themselves when suddenly faced with sliding sales and lay-offs.

The especially competitive nature of the industry encourages companies to indulge in marathon "pushes." These mammoth efforts usually involve close-knit teams of young engineers working night and day to solve a problem or to get a new product ready for launch. A particularly good account of such a push may be found in Tracy Kidder's book, *The Soul of a New Machine* (Little, Brown & Co, Boston, 1981). Kidder tells the true story of a one-year push to produce a new minicomputer, code-named "Eagle", at Data General Corporation, which was desperate to stay one jump ahead of the competition. Working all hours, the young team was pushed to the limits of its members' physical and intellectual endurance by its temperamental baby, which turned out to be the Eclipse MV/8000.

Because of the importance attached to work and an individual's work performance, Silicon Valley is much more of a meritocracy than most industries. Electronics engineers get promoted if they can deliver the goods, and to get the best out of people Silicon Valley firms have developed their own distinctive management style. This tends to be informal, flexible and egalitarian – the very opposite of the norms prevailing in traditional companies. Bureaucracy and petty rules are kept to a minimum. As firms grow, managers go to great lengths to sustain the entrepreneurial spirit in employees and to keep administrative units small through decentralization. And as we saw above, in order to attract and retain key personnel, companies allow employees to come and go as they please, to take the occasional paid sabbatical, and to use the company's free recreational facilities. Such a system works particularly well for gifted personnel, but it can be cruel on life's losers.

The human and social costs of the microchip are most evident at home in the family. Workaholic electronics engineers see little of their wives and even less of their children. Husbands find it hard to explain to their families what they are working on, and wives are sometimes barred for security reasons from even visiting their husbands' workplaces. Partly in consequence, the divorce rate in Santa Clara County is the highest in California, which itself has a divorce rate 20 percent above the nation's average. Further, there are worrying reports that the neglected children of electronics superstars are turning to alcohol and drugs in increasing numbers. The pressures can be particularly strong on the children in two-career households.

Rogers and Larsen say that the model 1950s employee described by William H. Whyte in his classic book, *The Organization Man* (Simon & Schuster, New York, 1956), has been superseded by "The High-Tech Entrepreneur." Organization Man worked for a large corporation or public sector bureaucracy. He kept regular 9-to-5 hours and was family-oriented, returning home for dinner every night with his wife and kids in his single-family dwelling in the suburbs. Organization Man was a contented conformist, conservatively dressed and dedicated to serving the organization. He did what he was told and went where he was sent.

"High-Tech Entrepreneur," on the other hand, is more likely to be divorced and living with someone else in a condo. He rarely sees his kids and doesn't give a damn about rules, dress or peer group pressure. This high-tech man is a high achiever; he prefers to work in a small outfit and shows no loyalty to any company, only himself. Indeed, he is secretly planning his own start-up, and confidently expects to retire a millionaire at the age of 40. Personal relationships and leisure do not much interest him.

The microchip industry has its fair share of colorful characters and flamboyant entrepreneurs like Jerry Sanders of AMD. These "superstars" love to flaunt their new-found wealth in the form of material possessions like white Rolls Royces, Ferraris and Porches – not forgetting their private planes, helicopters and luxury sailboats. And high-tech goodies like satellite TV dishes are themselves current status symbols in this high-tech culture. But critics point out that community life in Silicon Valley is virtually nil. There is an almost complete lack of interest in politics (although a strong tendency to hold unspoken right-wing Republican views) or religion, and a lack of concern is shown for local social, civic and charitable activities. In fact, Silicon Valley is something of a cultural desert. The only art exhibitions to be staged in recent years have tended to feature things like floppy disks and circuit boards. Outside of work, keeping fit is the main recreational activity. Some companies have even established special jogging trails to cope with the large number of daily runners. As one executive put it, "Everybody here is into peak individual performance."

As we saw in chapter 1, there have been many attempts to replicate Silicon Valley in the USA, Europe and Japan. But so far there is nothing to rival the original. There is – and probably always will be – only one Silicon Valley.

Whether the culture it has created will become the norm elsewhere is not entirely clear. Not everyone will be working in the microchip industry or even in associated industries like software. Silicon Valley is also very much a *Californian* phenomenon. Route 128 around Boston, for example, has a lot of high-tech companies, but it hasn't spawned a culture like Silicon Valley.

And yet, the types of things people are doing in Silicon Valley and the way they are going about them provide some useful pointers to future economic structures and cultural styles. Silicon Valley society has generated great wealth and has to its credit some remarkable technological achievements. But success has been marred by environmental pollution and social problems such as family disintegration, cultural sterility and persistent class inequalities. Similarly, the high-tech society of the future is likely to have its good and its bad side.

One thing is clear: in Silicon Valley there can be no going back to prunes.

4 The Telecommunications Explosion

Going Digital – Deregulation and After – The Global Marketing Battle – Fiber Optics – Cable TV – Satellite Communications – DBS – Cellular Radio – New Network Services: Electronic Mail, Videoconferencing – Videotex

On the day of the nuclear accident at Three Mile Island in 1979, the local phone company recorded 1,745,000 calls into and out of the stricken area. Without the use of telephones to coordinate the rescue effort, a serious accident could have turned into a major disaster like Chernobyl, and half of Pennsylvania might be uninhabitable now. If a similar emergency had occurred 160 years ago, for communication we would have had to rely on smoke signals, carrier pigeons or the Pony Express.

Today, we take the telephone for granted. Communicating is a normal human activity, and nothing seems more natural than picking up the phone and dialling your mother in Chicago. Americans do it more than three-quarters of a billion times per day. Telecommunication systems are the vital arteries of advanced industrial societies and without them things would pretty soon grind to a halt. Yet the telephone was patented only in 1876, in Boston, by Alexander Graham Bell (just a few hours earlier, it seems, than Elisha Gray). Now, little more than a century later, there are about 600 million telephones in use in the world, and we stand on the verge of a new and more far-reaching revolution in telecommunications.

Falling costs due to advances in microelectronics, fiber optics, software and digital technology and new techniques for collecting, storing, displaying and transmitting information – such as by cable and satellite – are bringing about the *convergence* of the computer, office products and telecommunications industries into what is now being called the "information-processing" industry. Put another way, because the transmission of data and the processing of data have become inseparable, the computer and telecommunications industries are meeting head-on in the office automation market. This has led to a dramatic industrial restructuring; computer companies now have to be into telecommunications, and telecommunications companies have to be into computers.

Increasingly, telecommunications systems are being transformed from people talking to each other to computers talking to each other. In the USA, for example, the local telephone service is growing by only 3–5 percent a year and long-distance calls by 14 percent a year. But the transmission of data is growing by 25–30 percent a year, according to a 1986 report from market researchers Frost and Sullivan. As William H. Springer of Ameritech says, "Until recently, the telephone network was something for voice-to-voice communication. But new technology is providing an array of possibilities for using the same network." Already there exists electronic mail, facsimile transmission, message services, mobile phones and videotex. The possibilities include routine teleconferencing, teleshopping and telecommuting.

Even the humble telephone set has not escaped. Once telephones were rented from the telephone company, lasted a lifetime, and came in just two styles and four colors. Now phones are sold almost anywhere and consumers are being urged to buy them in any variety of shapes, sizes and colors because they match the wallpaper, look cute or offer additional features. The supply of telephone handsets was liberalized in the USA in 1977 and in the UK in 1981, but it was not until 1983 that sales in the USA really took off. Even so, consumers soon tired of cheap, imported Mickey Mouse, Snoopy and ET phones, which had novelty value but little else. More popular were the more complex models which really did offer something different – such as cordless phones and feature-laden "executive" phones, which are a pointer to the computer terminal phone-of-the-future.

The key event in the recent history of telecommunications – and arguably the biggest thing to happen in the industry since the invention of the telephone – was the divestiture of American Telephone and Telegraph's long-standing monopoly over the US telephone network, announced on January 8, 1982, and implemented on January 1, 1984. The breaking-up of AT&T – or "Ma Bell," as it was called – into one smaller company and seven regional operating companies is by far the most important example of "deregulation" in the USA and has sparked a wave of "liberalization" and "privatization" throughout the Western world, especially in Britain, Japan, Australia and Canada.

Deregulation leads to increased competition, and this in turn has led to the internationalization of the telecommunications industry, now worth over $200 billion a year. New and surprising alliances between companies and countries have been formed. This major upheaval in an industry after decades during which the frontiers had been fixed has stimulated rapid growth, dramatic technological advances and a whole variety of new services. But it has also brought with it headaches for governments as they struggle to develop appropriate and coherent telecommunications policies for the twenty-first century.

The new telecommunications infrastructure now being built has been compared to the building of the railroads or the highway network – except that it will be completed in shorter time. It will cost huge sums of money and use massive amounts of materials, such as optical fibers and high-performance engineering plastics. Much of the current activity remains unseen, but its significance will become more apparent after 1990, when the deregulation dust has settled and people realize the true extent of the revolution wrought in the great 1980s telecommunications explosion.

Going Digital

The "convergence" of computers and telecommunications is occurring because telecommunications systems are going digital, thanks to developments in microelectronics. This means that data processing and data transmission equipment will be talking the same language – the language of the *binary code*. In the binary code, a low-voltage pulse equals "1," while its absence equals "0." Each number is a "bit" of information, and a stream of such numbers – a "bit stream" – can be transmitted at enormous speed over conventional wires and in vast quantity using high-capacity fiber optic cable.

Existing phone systems were built to carry the human voice and are therefore based on analog devices in which the message signal rises or falls in line with the original signal. But analog systems are slow, they have a low capacity, and translation is required before data can be transmitted on the network. In fact, a vast array of devices and services has been developed in recent decades to make machine noises palatable to the network. For example, *modems* provide MOdulation and DEModulation of digital signals into analog signals suitable for the phone network; *multiplexers* of various kinds collect signals from more than one slow-speed device and send them over a cheaper high-speed line; *network processors* switch signals or concentrate them once they are in the network; *packet-switching services* process digital signals into large chunks and send them quickly over rented telephone lines.

Digital exchanges have no need for such elaborate decoding equipment or services. Digital technology is fast, flexible, reliable – and is getting cheaper by the minute. Thus, digital exchanges and PBXs are replacing electromechanical systems, just as word processors are replacing typewriters, facsimile machines are replacing copiers and electronic storage systems are replacing filing cabinets. It is remarkable just how much can be carried on today's "intelligent" network, which has seen great technological advances in the last two decades. New services will continue to be developed and offered to the public. But ultimately, the real explosion in telecommunications will occur with the realization of the Integrated Services Digital Network (ISDN).

The ISDN will be able to handle all kinds of services that can be transmitted in digital form – voice, data, facsimile, even moving pictures. In effect, the telephone will become a computer terminal communicating with a huge range of sophisticated devices and services. ISDN will make possible interactive banking and retail services, give access to a variety of general and specialist databases and deliver information to your desktop terminal using new forms of electronic mail and electronic newspapers.

ISDN will operate on a worldwide basis, and remote access will mean that it won't matter where your terminal is located – it could be in the car or on the beach. In theory, a US salesman in Australia, for example, could call up his boss in California and discuss a graph of Far East sales displayed on his portable terminal, which is based on data provided directly from the company's European headquarters in Brussels. But there is concern that the absence of comprehensive agreements on international standards for ISDN could hold back progress, despite the efforts of CCITT, the International Telegraph and Telephone Consultative Council.

ISDN will not happen overnight. It will evolve over the next decade or so as more and more exchanges and phone lines are converted to digital technology. It will be a piecemeal process as operators install new switching facilities, transmission devices and local networks, and suppliers strive to achieve compatibility. To give some idea of the size of the task and the potential of ISDN, the number of business telephones in the USA has grown from 56 million in 1983 to 70 million in 1987. Yet over the same period the number of business computers, personal computers, word processors and terminals installed has grown from 20 million to over 60 million. That's an awful lot of hardware just waiting to be put in touch with other bits of hardware. Huge investment in new lines, new message switches and translation devices will be required. But sometime before the year 2000, ISDN will have become some sort of reality.

Characteristically, the Japanese plan to have the basics of their ISDN, called Information Network System (INS), in place by 1995 – at a cost of about $10 billion. The idea of INS was first announced by Dr Y. Kitahara, executive vice-president of the Japanese PTT, Nippon

Telegraph & Telephone (NTT), at a Third World Telecommunications Forum held in Geneva in September 1979. Kitahara described INS as the "marriage" of computers and telecommunications, and the Japanese expect it to be *the* key infrastructure of what they call the "advanced information society" of the twenty-first century.

Already, Mitaka-Musashino, a Tokyo suburb, has been chosen for an experiment with "telecommuting" or working at home in "satellite offices" on the prototype INS. Households have been provided with special digital phones, video terminals and other equipment like so-called "sketchphones," hooked up via digital exchanges and high-capacity optical fiber cables. The experiment is costing about $80 million.

The scale of the Japanese project would seem to dwarf anything happening elsewhere in the world, although the USA is more advanced in other respects. Prompted by the oil price crisis of 1973, the Japanese "targetted" information technology and the "knowledge industries." They see INS in particular as a way of saving on resources such as time, space and energy and transport costs. Summarizing the advantages of the digital network, Keiji Tachikawa, chief manager at the Plant Engineering Bureau of NTT, said that INS will provide "economical, integrated, efficient, flexible and high-quality communications for Japanese people."

The Europeans, on the whole, do not have such grand strategies, but Britain, Belgium, Holland, France, Italy, West Germany and Sweden are pressing ahead with the installation of all-electronic digital systems. Britain seems to be ahead, even though by mid-1986 only about 20 local exchanges using British Telecom's System X were fully in operation. But BT, which was sold to the private sector in November 1984, is offering many new services and is providing the basis for the evolution to ISDN through the installation of fiber optic cables between major centres and microwave transmission systems. An ISDN pilot service was started in June 1985, based on a digital exchange in the City of London. It was subsequently extended to customers in Manchester, Birmingham and Nottingham and will be available at 190 centers by the end of 1987, according to British Telecom.

BT's System X has been an abject failure in the international market compared with the successes of Sweden's L. M. Ericsson (the AXE system), France's CIT Alcatel (E 10) and Thomson CSF (MT), America's ITT (System 12) and Western Electric (ESS), Canada's Northern Telecom (DMS) and others. The poor record of System X suppliers such as STC, Plessey and GEC led BT first to drop STC and then actually to seek an alternative to System X among foreign companies, despite the Concorde-style development costs which some say ran to £1000 million.

One development that will aid the achievement of ISDN is the rapid proliferation of *local area networks* (LANs). LANs are basically high-speed wires that link terminals, mainframes, minis and personal computers in a building or on the factory floor for the purpose of swapping information. People on the "net" often share expensive add-ons such as storage disks and high-speed printers. There are over 50 different brands of net on the market, including Arcnet (Datapoint), Ethernet (Xerox), WangNet (Wang) and Z-Net (Zilog), as well as systems offered by AT&T and IBM. The remarkable and somewhat unexpected rise of the personal computer has given networking a big boost.

Most local networks are one of two kinds. "Star" networks have a central control unit, usually a PBX or mainframe computer, with devices coming off it in a star-shaped configuration. "Ring" or "bus" networks lack a central core, and the devices are joined together in a circular or ring configuration. The main reason for installing a network is to share data and programs, and most new PBXs are being made "intelligent" so they can handle data – these are called "smart" PBXs.

Deregulation and After

History records that it was Alexander Graham Bell who first demonstrated his "harmonic telegraph" on March 10, 1876. But it was Theodore N. Vail who built up AT&T's Bell System between 1878 and 1887 in the crucial years before Bell's last patent ran out on January 30, 1894. For very nearly the next 20 years, AT&T fought hard in the marketplace as competitors tried to get in on the act. But AT&T, bolstered by Vail's return to the helm after two decades in 1907, not only won out against the independents, but defeated nationalization proposals as well.

State-owned postal, telegraph and telephone authorities (PTTs) were being formed all over the world at this time: there was a widespread consensus that, because of the economies of scale involved in planning and operating a public network, considerations of efficiency made the telecommunications industry a "natural" monopoly to be entrusted to a single carrier. This led to pricing structures and cross-subsidization practices based on social objectives rather than strictly commercial ones.

On December 19, 1913, the US government settled the argument when AT&T agreed to submit itself to tight government regulation in return for a (private) monopoly over the US telephone network. That monopoly remained wholly intact until 1968 and mostly intact until

January 1, 1984. In the interim years, AT&T had grown into a corporate colossus. In 1929, for example, it became the first US company to generate revenues of $1 billion. By 1983 it had revenues of $65 billion. It also had 84 million customers, 3.2 million shareholders and nearly 1 million employees (a workforce second in size only to the US government). With assets of almost $150 billion – more than Exxon, Mobil and General Motors (the next three biggest) combined – AT&T was by far the largest privately owned company in the world.

Thus, for around seven decades, the US telephone system was quietly run as a public service by one company – a company that was regulated, but was basically untroubled by competition. Overseas the set-up was similar, although the state owned PTTs inhabited an even more sleepy and cozy world in which domestic monopoly suppliers grew fat and lazy on regular orders and guaranteed profit margins.

Yet, underneath the apparently serene surface, technology-driven forces were shaping a major realignment of the telecommunications industry. Changing technology, in the form of microelectronics, fiber optics, cellular radio and microwave satellites, was creating new products and new services as the computer and telecommunications industries converged. Aware of the growing demand for data transmission, AT&T itself needed to get into computing, while computer firms like IBM were demanding the right to participate in the telecommunications revolution. It was increasingly apparent to everyone that the old PTT-style monopolies were no longer appropriate to the modern world.

By the late 1960s, the liberalization of telecommunications was underway. In a pioneering decision, the Federal Communications Commission (FCC) in 1968 allowed a Texas company to sell its Carterfone, a non-AT&T device which connected mobile radios to AT&T's own lines. In 1969 the MCI Communications Corporation won the right to connect its own long-distance network to AT&T's local phone systems. Even so, by the end of the 1960s AT&T still controlled 90 percent of the US telephone service. Finally, on November 20, 1974, the US Justice Department filed an antitrust suit against AT&T, with the immediate purpose of breaking off Western Electric, the company's manufacturing division, from the rest of AT&T.

The suit dragged on and on. The first judge on the case died, there were lengthy pre-trial hearings, and it wasn't until January 1981 that the actual trial began. Meanwhile, the FCC had in 1979 allowed AT&T the right to sell non-regulated services such as data processing. Even as the trial got underway, there was a growing feeling among officials of AT&T and the government that the time had come to settle the case. Accordingly, on January 8, 1982, the world was stunned by the announcement that AT&T had agreed to be split up, although it was not until August 1983 that judge Harold H. Greene finally gave his seal of approval to the proposed divestiture.

Thus, on January 1, 1984, the "new" AT&T came into being, together with seven newly independent regional operating companies or "baby Bells" (see figure 4.1). *Nynex* takes in New York and six New England states; *Bell Atlantic*, based in Philadelphia, covers New Jersey, Pennsylvania, Delaware, Maryland, Virginia, West Virginia and Washington DC; *Bell South* of Atlanta takes in Georgia, North Carolina, South Carolina, Florida, Alabama, Kentucky, Louisiana, Mississippi and Tennessee; *Ameritech*, based in Chicago, serves Indiana, Michigan,

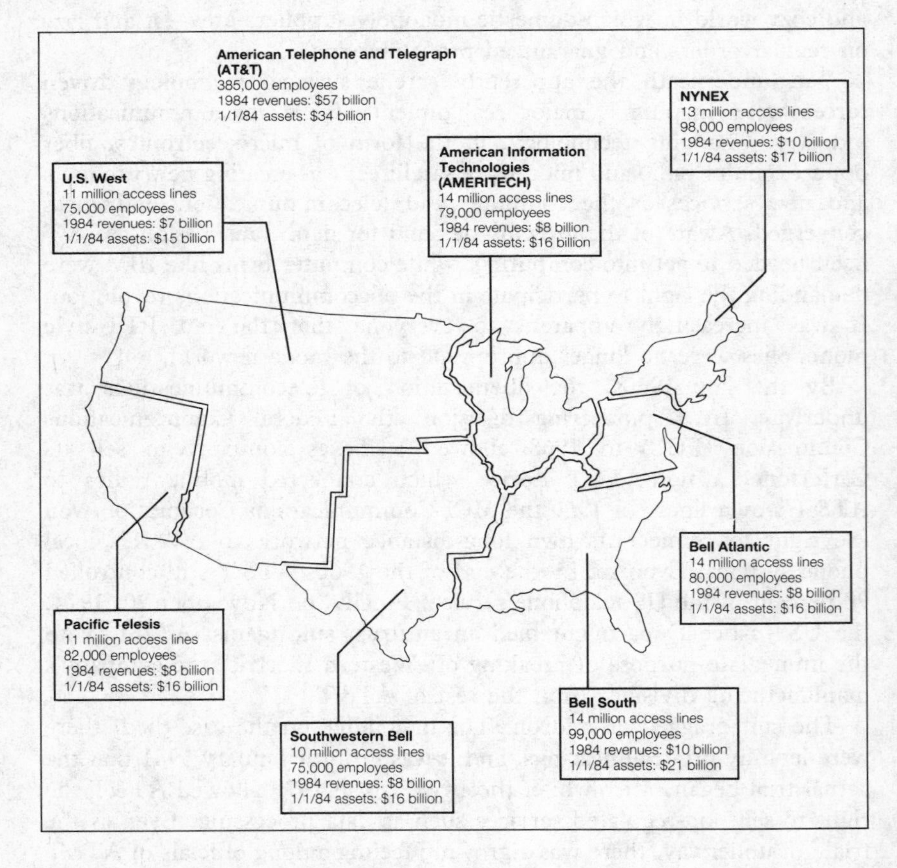

Figure 4.1 AT&T and the seven Baby Bells.
(Source: International Data Corporation)

Illinois, Ohio, and Wisconsin; *Southwestern Bell* of St Louis covers Arkansas, Kansas, Texas, Missouri and Oklahoma; *US West* of Denver takes the huge area of 14 states in the Midwest, Rocky Mountains and the Northwest; and *Pacific Telesis* of San Francisco looks after California and Nevada. Stockholders of the old AT&T with ten shares got ten shares in the new AT&T and one share in each of the seven regional companies.

The new AT&T is far smaller than the old AT&T. But with $35 billion in assets, it is still a giant – about the same size, in fact, as Mobil. It is made up of five parts: a long-distance phone service division; Western Electric, a telephone equipment manufacturer; Bell Labs, the famous research and development facility; an equipment and services marketing division; and AT&T International, which sells AT&T goods and services overseas. Even before divestiture day, AT&T International (founded in 1980) had struck a deal with the Dutch group Philips to market equipment jointly in Europe and elsewhere. But compared with ITT (founded in 1925), Sweden's L. M. Ericsson, West Germany's Siemens, Canada's Northern Telecom and Japan's NEC, AT&T is still a small player on the world stage.

Divestiture day did not settle things for good. Already, there has been a plethora of lawsuits and legal wrangles as the operating companies try to expand outside their territory and AT&T tries to become an operating company again. In August 1985, the Commerce Department's National Telecommunications and Information Administration (NTIA) recommended further deregulation, such as letting the "baby Bells" sell long-distance services after September 1986. The "baby Bells" were already branching out into everything from computer retailing, cellular radio and selling their expertise abroad to directory publishing and even real estate development (something not envisaged by Judge Greene), but they were *not* free to sell information services or long-distance services according to the terms of the divestiture judgment.

In September 1985, the FCC gave AT&T permission to market office automation and computer services in conjunction with its telephone services (again, something the divestiture judgment had forbidden it to do). In February 1986, Judge Harold H. Greene himself gave Pacific Telesis permission to operate outside its area – by allowing it to purchase Dallas-based Communications Industries Inc., which provides paging and mobile telephone services across the country. But he stopped Bell Atlantic from undertaking a similar expansion on the grounds that they hadn't asked permission first!

Customer reaction to divestiture has not been favorable. While AT&T and the regional phone companies argue over charges, the use of lines and the responsibility for repairs, customers complain of poor service and long waits for the installation of new lines. Obviously, problems were only to be expected with such a massive reorganization, but there are those who argue that the level of service may never be the same again. Other drawbacks include the loss of one-stop telephone service, higher residential phone rates, higher charges for equipment installation and more customer confusion over telephone bills and the various new "options" available.

On the plus side, competition has led to the faster introduction of new technologies and services, lower long-distance call rates, easier access to long-distance discount carriers, lower equipment costs for the phone companies and the more equitable pricing of the phone service with the elimination of cross-subsidization.

For 15 years prior to deregulation, new networks had been nibbling away at AT&T's monopoly. Their origins go back to the immediate postwar period, when companies started to experiment with their own microwave radio networks. But it was not until the late 1960s and early 1970s, as we saw above, that the FCC gave the independents permission to operate. MCI was the pioneer of alternative long-distance networks, (see Larry Kahaner's history of MCI, *On the Line*, Warner Books, New York, 1986), while GTE Sprint began with cables run along the Southern Pacific Railroad's rights of way. The new networks achieved spectacular growth after the resale of circuits became legal in 1978.

In 1980 Allnet Communication Services came into existence, buying circuits from the independents themselves and reselling them – the same game the MCIs and GTEs had been playing with AT&T. In addition, United Telecom, Western Union, RCA and ITT all offer long-distance and international telecommunications services. Satellite-based networks are also operated by most of the independents, especially American Satellite, GTE (Telenet), Satellite Business Systems (SBS), Tymshare, (Tymnet), Hughes and Vitalink.

From divestiture day on, an all-out battle for long-distance traffic developed between these alternative carriers and AT&T. In the divestiture agreement, Judge Greene ruled that local phone companies had to provide "equal access" for customers to alternative networks by September 1, 1986. Individual subscribers would be able freely to elect a long-distance carrier as the new system was phased in. After September 1, 1986, subscribers wishing to change their carrier would have to go through a complicated procedure and pay a fee. But it was also agreed that the 55 percent discount on local connections enjoyed by the alternative carriers would be phased out at the same time: they would have to pay the same as AT&T. In order to meet the new competition,

AT&T made progressive cuts in its long-distance rates so that by mid-1986 they were 19 percent lower than they had been on divestiture day.

With AT&T still holding more than 90 percent of the long-distance market in early 1985, the FCC decided in June 1985 that customers who failed to specify a long-distance carrier would be assigned a carrier in direct proportion to those who *had* made a choice, instead of letting them all go to AT&T. This was a victory for the independents, but there was growing concern among them over what was actually meant by "equal access": in practice, most local phone companies were converting to the new system on a piecemeal neighborhood-by-neighborhood basis, which failed to offer subscribers a real choice. With such a high proportion of existing subscribers, AT&T had the inside track in terms of customer information, while its cut-rate competitors had to rely on general mail shots which couldn't be precisely targeted.

As the September 1986 deadline loomed and the last 30 million out of 86 million individual subscribers made their choice, the great long-distance telephone election reached its frenzied finale. With a $46 billion market at stake, AT&T and its rivals poured over $500 million into a late advertising blitz (AT&T's successful TV ads featured an effective performance by Cliff Robertson – MCI responded by hiring Joan Rivers and Burt Lancaster). As AT&T's market share drifted down below 80 percent, MCI (by then 30 percent owned by IBM and merged with SBS) achieved a large enough market share to remain viable. But there was a big question mark over the future of GTE Sprint, despite its merger with US Telecom. Most analysts were predicting that only two major carriers would remain in the long-distance market in the long term.

Another sign of the telecommunications explosion is the way some major corporations and institutions are literally setting up their own phone networks. The problem is that local phone services (or "loops") cannot cope with the masses of data people now wish to transmit. High-speed data bounced off satellites has to crawl "the last mile" into the office. However, by installing their own microwave dishes and pointing them at satellites or other buildings, companies can "bypass" the local phone system altogether. Improved technology is making this much easier to do, and soaring local phone rates are providing the incentive. Since a very high proportion of their business comes from just a few big customers, metropolitan phone companies are extremely worried about the growth of "bypassing." And it will inevitably lead to increased charges for residential customers.

The Port Authority of New York, for example, is building a giant $84 million Teleport on Staten Island. The idea is to bundle together enough long-distance voice, data and video traffic to get bulk discounts on satellite circuits. The New York Teleport will have 17 earth stations beamed at all major commercial satellites, and its fiber optic cables will be able to carry 38.8 billion bits of data per second – or 300,000 simultaneous phone calls – into Manhattan. Backed by Merrill Lynch, Western Union and the Port of New Jersey, the project is completely independent of the New York Telephone Company. More than 20 other teleports are planned for major US cities, and their backers see them as providing the new communications gateways to the business districts of metropolitan areas, just as airports and railroad stations are the transportation gateways.

The Global Marketing Battle

Deregulation in the USA has been accompanied by deregulation in overseas markets. In Britain, the countdown to liberalization and then privatization of the publicly owned monopoly over the UK telephone service began with the newly elected Conservative government's decision in September 1979 to split the Post Office into two separate divisions: post and telecommunications. This came about in early 1981. In July 1980 the British government ended the Post Office's long-standing monopoly over the supply of telephone equipment, such as phones, answering machines, telexes and PBXs. And in September 1980, Professor Michael Beesley of the London Business School was commissioned to report on the proposed liberalization of the use of the UK telephone network.

The Beesley Report, published in April 1981, came down firmly in favor of opening up the network to operators who wished to provide new services to third parties. Meanwhile, the government announced that the private maintenance of some PBXs would be allowed and the old colonial telecommunications company, Cable & Wireless, was to be sold to the private sector. In July 1981 the UK government decided to accept the recommendations of the Beesley Report, despite objections from the management of the newly formed British Telecom (BT).

By early 1982 there was talk of privatizing BT, and Mercury Communications, a new consortium made up of Cable & Wireless, BP and Barclays Bank, was granted a license to provide an alternative network to BT using fiber optic cables run along British Rail lines. In July 1982 the British government announced the proposed sale of BT in a White Paper (Cmnd 8610) entitled *The Future of Telecommunications in Britain*. BT was valued at about £8 billion, and potential investors would get a company with 20 million customers and 240,000 employees. Fierce debate over the proposals followed, with strong opposition (including industrial action) being expressed by the BT unions. A Telecommunications Bill giving effect to privatization was promoted in November 1982, and meanwhile Professor Stephen Littlechild of Birmingham University was asked to report on ways of ensuring that a privately owned BT wouldn't exploit its dominant position in the market.

Published in January 1983, the Littlechild Report emphasized competition rather than regulation and suggested ways in which the network could be kept open for independents. While BT itself was being slimmed down and "readied for the market" by chairman Sir George Jefferson, the Telecommunications Bill passed through Parliament despite fierce opposition and after a short break for the June 1983 general election. It finally became law on August 6, 1984. British Telecom was to be sold, but it would have to adhere to the terms of its license and be closely watched over by a body called Oftel (or the Office of Telecommunications) under the direction of Professor Bryan Carsberg. Mercury – now wholly owned by Cable & Wireless – was also granted a license.

Amid considerable publicity, and after a massive advertising campaign, the sale of 51 percent of the shares in British Telecom for £3.5 billion was accomplished in November 1984. For a staid, slow-moving, traditionalist country like Britain to have carried through such a major reorganization of something so basic as telecommunications in a fairly short space of time was a remarkable achievement. Even so, telecommunications in Britain has been deregulated only to a degree – the direct resale of circuits, for example, is *not* allowed and has been expressly ruled out until at least July 1989.

Moreover, during 1985–6 steps had to be taken by Oftel to safeguard competition by bolstering Mercury's position and reducing BT's dominance. Early in 1985, Mercury went to the High Court to force BT to let it connect customers to the Mercury network on BT's local circuits. The matter was taken up by Oftel, which ruled in Mercury's favour in October 1985 by allowing full interconnection. This ruling guaranteed Mercury's existence and meant it could now offer a complete alternative phone service. It also led directly to BT reducing its charges for long-distance calls and, as in the USA, increasing its charges for residential subscribers (subject of another critical Oftel review).

In addition, BT was forced to adopt a code of practice ensuring confidentiality of customer information, and there was a further row over BT's charges for private leased circuits. In early 1986, BT was urged by Oftel to set up a separate office equipment and PBX subsidiary, and BT lost its monopoly over installing domestic telephone wiring. On May 15, 1986, Mercury started its alternative phone service – in the first instance, for business customers only. Mercury also did a deal with AT&T to take its international calls to the USA.

Meanwhile, liberalization of the telecommunications market was proceeding apace. BT started buying exchange equipment abroad from AT&T, Philips and especially Thorn-Ericsson, a Thorn EMI–L. M. Ericsson joint venture. BT's purchase of ailing Canadian PBX manufacturer Mitel was referred to Oftel and the Monopolies and Mergers Commission, but it was given the go-ahead by the UK government in January 1986.

In Japan, the Nippon Telegraph and Telephone Public Corporation (NTT), modelled on AT&T when it was founded in 1952, has also been privatized, and along British lines. The initial plan, conceived by chairman Mr Hisashi Shinto in 1982, was for NTT to be split up into one central company and five regional companies along the lines of the AT&T divestiture. However, after studying the plans for BT, Shinto and the Japanese government decided simply to privatize NTT and allow competition to spread inwards from the boundaries. But unlike BT, the privatization of NTT and the liberalization of the Japanese telecommunications market – which took effect on April 1, 1985 – is proceeding more gradually.

With 60 million sets and 42 million lines, NTT is much bigger than BT (24 million sets, 20 million lines), and it is therefore worth a lot more. In Japan, as in Britain, the government believes that the old arthritic state bureaucracies are not capable of coping with the demands of the technological revolution. Through 1985 and 1986, the momentum of privatization was sustained by a series of diversifying moves by NTT. But there was growing tension between the US and Japanese governments over the slow pace of liberalization of the Japanese telecommunications market, which many saw as a crucial indicator of Japan's willingness to do something about trade friction in general.

Further liberalization and privatization elsewhere seems inevitable – but progress in continental Europe, for example, has been slow. Belgium began a process of liberalization of its telecommunications market in 1984–5, and the Dutch decided on deregulation (but not privatization) of their state PTT monopoly in 1985. There has also been a slight loosening of regulation in Spain, Italy and Norway. Even so, the anti-deregulation forces – strongest in West Germany (and France before the 1986 elections) – have remained powerful, despite a chorus of complaints from business users (especially US businessmen) frustrated by waiting lists for equipment, repair delays, high charges, low capacity and the lack of Europe-wide standards.

The US government is doing its best to export the new competitive ethos abroad. The belief is that consumers around the world will demand the best and latest in technology and that competition is better able to provide the goods and services required by the evolving information society. PTTs that stay regulated will fall behind in technology, and nations concerned will have inferior infrastructures. Gaining market access has therefore become one of the basic forces in the realignment of the telecommunications industry, along with the search for economies of scale as the cost of developing digital exchange equipment soars (so obvious, it's often overlooked) and the need for mastering the different technologies required in the new hybrid telecommunications products demanded by an increasingly sophisticated market.

As State Department officials spread the word at international conferences and in bilateral trade talks, US corporations are moving into newly liberated foreign markets by setting up subsidiaries, buying into domestic suppliers or entering into alliances and technology-sharing agreements so they can offer a complete range of equipment for specific domestic markets. The best example is AT&T, which has joint ventures going in Taiwan, Singapore and South Korea. In Europe, AT&T has already teamed up with Italy's office equipment marker, Olivetti, the Dutch electronics group, Philips, and the French switch-maker, CGE, to offer a greater range of products. It also has high hopes in Spain. In addition, AT&T has links with no less than 16 Japanese companies seeking to take advantage of liberalization. As *Fortune* magazine put it, these deals "will help the company to blend in with the locals."

The global marketing battle now underway is for very high stakes. Only about 30 percent of the world telecommunications market is open for competition (half of it comprises the USA), but even so it is worth billions of dollars. Growth in all markets will be dramatic in the next two decades, especially in Asia and the Third World countries. There has been tremendous competition for "turnkey" projects in certain countries like India and particularly China, with the US and European governments offering all kinds of bribes and incentives on behalf of their domestic suppliers.

Much of this action so far has been in the USA itself. ITT, the world's second largest telecommunications equipment maker, was fighting hard to re-enter the US market, while Canada's Northern Telecom and Mitel grabbed an impressive slice of the US switchboard market in a short space of time. In the highly competitive exchange equipment market,

Japan's NEC and European firms like Sweden's L. M. Ericsson (teamed with Honeywell), France's CIT–Alcatel, Britain's Plessey (which purchased Stromberg–Carlson in 1982) and West Germany's Siemens have piled in – with Siemens doing especially well in getting business from the "baby Bells." But ITT had to pull out of this particular race in early 1986, having spent $150 million in a vain attempt to adapt its System 12 digital exchange equipment to the US market.

In Europe, all kinds of pacts and alliances have been formed. In France CIT–Alcatel and Thomson CSF have teamed up, and they have links with Italy's state-owned Italtel, which itself has joined with the Italian firm Telettra and America's GTE International in a new venture called Italcom. In addition to those mentioned above, there are many other marketing alliances blossoming in the new areas such as fiber optics (Corning Glass, for example, has links with Siemens, Thomson–CSF and Pirelli), videotex (for example, IBM and West Germany's Bundespost) and cellular radio (for example, Plessey, Stromberg–Carlson and Mitsubishi). These technology-sharing agreements are usually for the specific purpose of selling certain kinds of equipment in the different European markets. Such arrangements help guide overseas suppliers through the maze of local PTT regulations and market conditions.

Much of this hectic European activity has been stimulated by fear of IBM. International Business Machines – or "Big Blue," as it is known in the trade – has in various ways signalled its intention to enter the expanding telecommunications market. This became obvious after the US government in 1982 dropped its antitrust charges against it (first filed in 1969), and certain after the divestiture of AT&T. IBM owns one-third of Satellite Business Systems and 30 percent of MCI; it has a significant presence in local area networks and is involved in videotex developments. IBM has also refined its "systems network architecture" (SNA), a series of protocols that allow previously incompatible computers and peripherals to talk to each other. Perhaps most important of all, IBM's outright purchase of PBX maker Rolm in October 1984 (after first buying a small stake in June 1983) means that it now has an office switchboard with which to spearhead its attack on telecommunciations markets – especially Europe.

European manufacturers now have less than a fifth of the total European market for information-processing goods, and there is a general belief that Europe has left it too late to get its act together. France's Bull, Britain's ICL and West Germany's Siemens have set up joint research projects, while the European Community agreed to fund ESPRIT and RACE, two cooperative research programs. But such collaboration tends to be the exception rather than the rule, and

European production remains fragmented: there are still about ten European producers of telephone exchange equipment. Even joint projects between firms of the same nation (like the ill-fated tie-up between Britain's GEC, Plessey and STC to sell System X) tend to fall apart, and the main axis for European reorganization is more likely to be transatlantic.

Meanwhile, US companies like ITT are establishing strong positions with their wholly owned and partly owned subsidiaries in major European countries (like STC in the UK, which in 1984 took over ICL, Britain's only mainframe computer maker). IBM has reached agreement with West Germany's Bundespost over SNA and is building extensive links with British Telecom, under the watchful eye of the UK government. A long-awaited anti-monopoly suit filed against IBM by the European Commission in 1980 alleging abuses of its dominant position in mainframe computers (IBM has 80 percent of the European market) was settled in July 1984, when IBM agreed to modify its products to conform with EEC standards and the EEC agreed not to force disclosure of technical details.

Many observers are now predicting that the main battle for the information-processing market in the next decade will be between AT&T and IBM – symbolized, it is said, by their adjacent blocks on Madison Avenue in midtown Manhattan. Thought by many to be losing its touch, IBM has become increasingly aggressive in recent years, especially in Europe. IBM perceives that transporting data between computers will eventually be nearly as important as manipulating data in a mainframe or personal computer. It therefore intends to maintain and extend its position in computers and office equipment by offering customers the communications equipment needed for a total system. AT&T, despite its strong position in telecommunications equipment via Western Electric, faces what is probably a much tougher task of getting into IBM's business of selling computers – but it does have Bell Labs and a lot of money behind it. If IBM wins out, we will have to conclude that deregulation not only meant the breaking-up of "Ma Bell," but also paved the way for the transformation of "Big Blue" into "Ma Blue."

Fiber Optics

One of the new technologies transforming telecommunications – and perhaps now the most important one – is a remarkable innovation, fiber optics. Although various people including Alexander Graham Bell experimented with using light to transmit speech 100 years ago, modern optical fibers date from 1966, when two British engineers at Standard Telecommunications Laboratories in Britain first demonstrated their potential. However, it was Corning Glass Works of Corning, New York, that produced the first commercial version in 1970, and latterly it has been Japanese companies like Sumitomo that have exploited the invention most dramatically on world markets.

Optical fibers are tiny strands of pure glass no wider than a human hair that can carry thousands of telephone conversations or other digitized data in the form of extremely fast streams of light pulses. Many such strands can be bundled together into a cable, which is typically about one-fifth of the size of a conventional copper cable. Light travels down the strands and is converted back into sound or whatever at the other end.

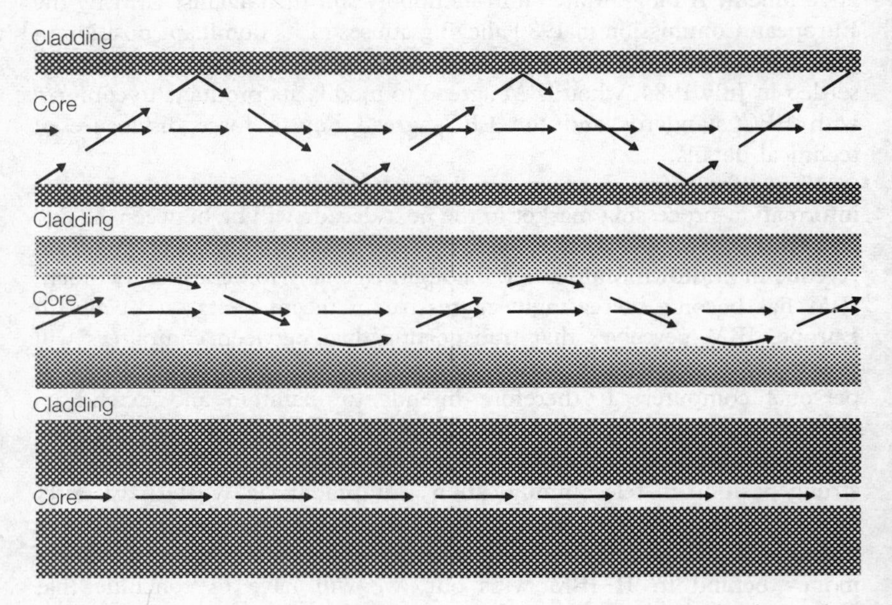

Figure 4.2 The three fibers of fiber optics. "Step-index" fiber (top) consists of glasses of two different densities, called core and cladding. Light zigzags down the core, bouncing off one side and then the other of the core-cladding interface. In "graded-index" fiber (center), the glass in the core varies in density, so the light travels in a smooth, curving path. Thus, its information content is less distorted. In "monomode fibers" (bottom), the core is very small in relation to the wavelength of the transmitted light. The light therefore moves down the fiber in a straight line with very low distortion.
(Source: "Optical Fibers: Where Light Outperforms Electrons" by Les C. Gunderson and Donald B. Keck, *Technology Review* May/June 1983, p. 37. Copyright, 1983 Bell Telephone Laboratories Inc., reprinted by courtesy of AT&T Bell Laboratories.)

There are basically three kinds of optical fibers (see figure 4.2). The simplest is "step-index" fiber, in which light zig-zags down the central core, bouncing off the sides which are covered with reflective cladding. In "graded index" fiber, the glass in the core varies in density toward the edges, so the light travels in a smooth, curving path. Finally, in single or "monomode" fiber the core is very small in relation to the wavelength of transmitted light, which therefore travels almost in a straight line.

Optical fibers have a number of advantages over conventional copper cable, which is costly, slow and has a low capacity. Optical fibers have a much higher capacity – a single strand can carry as many as 2000 simultaneous telephone conversations – and they are much faster. They are also becoming cheaper to produce because copper is more expensive than sand, the main ingredient of glass: in 1977 a meter of fiber cost $3.50; in 1985 it cost $0.25.

Thus, optical fiber is quickly becoming the preferred medium for transmitting voice, data and video, especially over long distances. Telecommunications applications account for more than 60 percent of the optical fiber market. According to market researchers Frost and Sullivan, sales of optical communications equipment will leap from $440 million in 1985 to $1.2 billion in 1989.

Optical fibers are easier to handle because they are thinner and lighter and thus fit more easily into crowded underground ducts. They are also easier to use because fewer "repeaters" are required on the line to boost the signal, and the few that are needed can be installed in warm and dry telephone exchanges rather than damp holes in the ground. Optical fibers are free from the kind of electrical and environmental interference that plagues satellite communications, and they offer greater security because they cannot easily be "tapped." They are particularly suitable for applications in defense, airplanes and automobiles and for use in hazardous places, although recent British research has raised some doubts about their durability. But clearly, they have a lot going for them.

The glass used in fiber optics has to be extremely pure. Two things can cause light to be lost and thus transmission impeded: "absorption" of light, which is caused by impurities in the glass as few as even one or two parts per million, and "scattering," which occurs when the light wave is deflected by even quite small changes in the density of the glass medium. The breakthrough at Corning in 1970 came with the demonstration that it was possible to produce very pure glass of uniform quality suitable for transmitting light over long distances. In the production process invented at Corning – known as "vapor deposition" – high-purity vapors of silicon and germanium are reacted by applying heat to produce layers of glass. The composition of the glass can be altered by changing the proportions of the ingredients to produce the different densities required for graded-index fibers.

In 1977, scientists at NTT's Ibaraki laboratory in Japan came up with an improved version of vapor deposition known as VAD (for Vapor phase Axial Deposition). Now, three of Japan's leading optical fiber makers – Sumitomo, Furukawa and Fujikura – have developed a new technique which makes possible the mass production of optical fiber by VAD: up to 20 km of thin filament can be drawn from one end of a hot glass rod, compared with 5 km using conventional techniques. It is said that, if the oceans were as transparent as the glass in state-of-the-art optical fibers, it would be possible to see the bottom of the Mariana Trench from the surface of the Pacific – 32,177 ft down!

The process of transmission used in optical fibers is called pulse code modulation and is rather like a fancy version of Morse code. The light pulses are originated by laser diodes which are capable of millions of blinks per second. Repeaters boost the signal if necessary, and a photo-detector receives the signals and converts them back into their original form at the other end. Today, the fastest commercial lasers in use can pulse at around 400 million times a second, enough to transmit 6000 phone calls along a single optical fiber. That means that the equivalent of the entire contents of the *Encyclopedia Britannica* can be transmitted in about 8 seconds. At Bell Labs and in Japan, lasers have been demonstrated which can send 2 billion pulses each second over a distance of 40 miles; sending the same amount of information over a conventional telephone line would take 21 hours. And there is talk of ultra-high speed soliton lasers that blink in picoseconds – or a million-millionth of a second (10^{-12} of a second).

Progress with installing fiber optic cables is now very swift. From just 3700 miles of cable installed worldwide in 1980, the total mileage is expected to leap from around 40,000 in 1986 to over 100,000 by 1989. In 1983 AT&T announced a plan to link ten cities in the USA from Boston to Richmond, Virginia, with 1200 miles of cable, and it expects that by the mid-1990s all major exchanges will be served by optical fiber cables. By the year 2000, AT&T predicts, copper cable will be a thing of the past.

MCI, the independent carrier, is installing 1600 miles of cable, much of it along the Amtrak railroad between New York and Washington. Another railroad, CSX Inc., is planning to lay no less than 5000 miles of cable in partnership with the Southern New England Telephone Co. called the LightNet network. But the grandiose plans of Fibertrak to install 8000 miles of cable along the tracks of the Santa Fe Southern Pacific and Norfolk Southern Railroads had to be postponed in 1985.

Meanwhile, work has started on the first transatlantic submarine fiber optic cable, called TAT-8. AT&T is in charge of the biggest section, but the British and French are also building sections of this $335 million link scheduled for completion in 1988. TAT-8 uses a 1.3 micrometer wavelength system which can carry light pulses for a much longer distance and therefore requires fewer repeaters (1.3 micrometer systems might be superceded by the even better 1.55 micrometer systems). Also in the transatlantic race are Tele-Optik, a joint venture between Britain's Cable and Wireless and E. F. Hutton; their PTAT-1 cable should be finished by 1989. At about the same time, Submarine Lightwave, founded by former Western Union International executives, hopes to complete its TAV-1 transatlantic cable.

The transpacific submarine optic cable race is between an international consortium headed by AT&T, which is building a 7200-mile network called TRANSPAC-3, and HAWAII-4, which will link California with Hawaii and then split into two branches – one going to Guam, the other to Japan. A further link will be made to the Australian Overseas Telecommunications Commission's undersea fiber optic cable network, which will initially link Australia and New Zealand across the Tasman sea. TASMAN-2, as it is called, will then travel north to link with Japan and the USA. The first, trans-Tasman, phase of the project, which will use a 1.5 micrometer system (requiring repeater stations only every 150 kilometers), is due for completion in 1991. Some 60 percent of Australia's overseas communications are currently carried by satellite, but that is expected to be reversed as fiber optic cable takes over.

The British, who helped pioneer the use of optical fibers in telecommunications, have laid a submarine cable, called UK–Belgium 5, under the North Sea to link Britain with Belgium. It will require only three repeaters, but they will have to be buried in order to avoid damage from French beam trawlers. (The proposed Channel Tunnel could solve that problem at a stroke.) British Telecom was in fact the first PTT to commit itself to optical fiber in 1981, and BT has a local network based on optical fibers operating in the new city of Milton Keynes, in Buckinghamshire. By 1990 over half of BT's 28,500-mile trunk network will use optical fiber. In addition, the independent UK carrier, Mercury, has based its network on optical fiber cable run along British Rail tracks. The UK is Europe's largest market for optical fiber.

The French have ambitious plans to cable the whole of their country with optical fibers. But the French government has a habit of announcing grandiose schemes which never come to fruition, and already the re-cabling project has been held up by political wrangling between DGT (the French PTT), the government and local authorities.

Agreement was finally reached in April 1984, but many doubt whether DGT and its suppliers CIT–Alcatel and CGCT (the nationalized former subsidiary of ITT) will be able to meet the target of 1 million homes cabled each year and 6 million by 1992.

In West Germany, where telecommunications are also firmly under the control of the state-owned PTT, they are spending $300 million a year under the BIGFON program to wire up major cities. About 2.5 million households now have access to cable TV. Meanwhile, the Dutch maintain their traditional lead in cable TV based on two decades of experience; in Holland no less than 85 percent of homes have been connected – the biggest copper coaxial network in Europe. The Dutch government has also launched two local interactive TV experiments, in Geldrop, near Eindhoven, and Limburg – although only the former uses optical fibers.

The Japanese aim to be the first country in the world to run optical fiber cables into every home. Over the next 15 years, NTT plan to spend a massive $80 billion on re-cabling, starting with the 1800-mile Tokyo–Hiroshima–Fukuoka–Sapporo link, which forms the backbone of the Japanese information network system (INS). The Japanese targeted fiber optics a decade ago, building up production capacity (set to increase six-fold by 1990) and cultivating markets. There is also a long-running two-way interactive TV experiment at Higashi–Ikoma, near Osaka. Bolstered by huge spending by NTT in the domestic market, Japanese firms such as Fujitsu, NEC and Sumitomo have invaded the USA, directly supplying independents like MCI and GTE Sprint and building their own plants in Texas, North Carolina and Virginia.

"Opto-electronics" – the term covers a range of devices which convert light to electric current and vice versa – and optical fibers in particular have enormous potential. The Japanese realized this early on, and in 1981 Japan's Ministry of International Trade and Industry (MITI) started its Opto-electronics Joint Research Laboratory, a well funded program linking a number of companies and aiming to develop optical integrated circuits. The idea behind it is that optical fibers will not only replace conventional wiring, but also, along with semiconductor lasers (or laser diodes), will form the basis of a new kind of computing – one based on integrated opto-electronics rather than on transistors in semiconductor chips, as in today's computer technology. Optical processing would be much faster than electrical processing, and optical chips would contain many more components than even the current generation of VLSI chips. Hybrid devices containing both electrical and optical circuits have already been devised by the Japanese, and some analysts are even suggesting that we will see an "optical computer" before the year 2000.

Cable TV

The invention of optical fibers means that it is now theoretically possible to "wire" the whole of society; one fiber optic cable into each home would have the capacity to provide a telephone service, numerous TV channels, two-way interaction and rapid data transmission, as well as a host of new facilities such as videophones, teleshopping, home banking and telecommuting. There have been many visionary calls for the "Wiring of America": for example, in *The Wired Society* (Prentice-Hall, Englewood Cliffs, NJ, 1978), prolific author James Martin describes cable networks as the new electronic "highways" of the future. In the UK, especially around 1982, there was much talk of the "Wiring of Britain" and the creation of a "National Electronic Grid" (Lord Weinstock). Britain's Information Technology minister at the time, Kenneth Baker, likened the coming of cable to the construction of the railroads and the telephone network. Cable represented the "third communications revolution," he said. However, despite the euphoria that hit the USA in 1980–1 and the UK in 1981–2, the dream of the Wired Society is not much closer to becoming reality.

Even basic cable TV has had its ups and downs. Despite phenomenal growth in the USA, few cable TV networks are making money and there is considerable uncertainty about the future. Cable TV networks for the most part do not use fiber optic technology (except for some trunk lines), but they are extensive: starting out in 1948 as a way of improving reception in isolated areas, the US cable networks had 1 million subscribers by 1965. Then two decades of rapid expansion followed: there were 4 million subscribers by 1969, 11 million in 1977, 16 million in 1979 and 33 million in 1984. Over 40 percent of US households now have access to cable.

The mid-1970s surge coincided with the introduction of Time Inc.'s Home Box Office and other channels which offered home viewing of first-run, uncut feature films. A profusion of networks and different kinds of programming rapidly developed, with most networks offering "basic" or "first-tier" services resembling conventional TV. The "second tier" is really pay-TV – viewers have to pay an additional fee to see live concerts, sports events and feature films. A "third tier" of two-way services is also available in a few local areas, such as Columbus, Ohio, where 38,000 residents participated in the Warner–Amex QUBE interactive TV experiment.

The success of Home Box Office, the pop music channel MTV and some sports channels led to a veritable cable boom in 1980–1. Operators fell over themselves to sign deals with local communities, who were able to extract all kinds of fees and concessions in return for cabling rights – including the construction of public-access TV studios and, in one case (Sacramento), the promise by the cable company to plant 20,000 trees in local parks. But even as early as 1982 doubts began to creep in about the economics of cable, and there were reports that most cable companies were making heavy losses – squeezed by rising costs and earnings well below expectations. In September 1982, CBS's prestigious cultural channel, CBS Cable, folded, and this was rapidly followed by more failures.

By late 1983, the scramble to get into cable had turned into a wholesale retreat. Construction costs and interest rates were soaring, cable companies found they were locked into impossible contracts with city governments, and the inevitable shake-out occurred. Merger fever ran high: for example, Ted Turner's Cable News Network, based at his WTBS "superstation" in Atlanta, bought out its Viacom–Westinghouse rival, Satellite News Channel, for $25 million. In fact, only WTBS, CBN Cable Network and the Christian Broadcasting Service were reporting profits. In addition, ratings were not growing as expected and as a result advertising revenue was too low to impress backers. Indicative of the crisis in cable, Time Inc.'s new magazine, *TV-Cable Week*, folded after only six months with losses totalling $47 million (see Christopher M. Byron's book, *The Fanciest Dive*, W. W. Norton, New York, 1986).

Despite these serious problems, new subscribers were still being signed up at the rate of 300,000–400,000 per month in 1983. But this fell back to about 200,000 per month in early 1984, and even the highly successful Home Box Office (with 14 million subscribers) saw a sudden slowdown in its growth. Forecasts of 46 million subscribers by 1988 were revised downward to 43 million or less, and it appeared that the great cable TV boom was over. Some lay the blame for the slump on rising sales of VCRs, while others speculated that subscribers were tiring of the repetitive nature of the programming. There was also a shortage of new movies.

Even so, some companies like TCI and ATC seemed to be doing all right, and analysts suggested that this was because they and others had stuck to the basics of cable – the provision of news, sport and entertainment; it was companies like Warner–Amex, which were attempting something more ambitious, that were running into financial trouble. A "new realism" or "back-to-basics" trend seems to be pervading the industry. Cable TV will survive, but it now appears less likely that cable will become an alternative communications system or will provide those electronic information highways of the future – not for a while, anyway.

The UK has had a similar experience with cable – of initial euphoria and boom turning to gloom – but on a different scale. Like the USA, cable was used from 1951 onwards to distribute TV broadcasts in areas where reception was poor. By 1982 around 2.5 million households had been cabled and there had been a handful of local "pay-TV" experiments. But, unlike the USA, broadcasting has always been more tightly regulated in Britain (through the BBC–IBA duopoly), and the prospects for cable in Britain have been inextricably linked with (changing) plans to legislate for direct satellite broadcasting (DBS).

In March 1982, the prime minister's recently formed Information Technology Advisory Panel (ITAP) delivered its report on the potential for cable TV in Britain. ITAP was full of enthusiasm: it recommended 30-channel two-way cable TV and envisaged the wiring up of half of all UK households by 1986! The Panel's basic philosophy was that privately funded (unlike most of Europe) entertainment channels would pioneer the wiring of Britain and pay for all the other goodies to follow. But, strangely, it advocated the continued use of copper coaxial cables for all but the trunk network, ruling out higher-capacity optical fiber on the grounds of cost.

Cynics suggested, with some justification, that ITAP was little more than a lobby for existing cable operators, who were seeing a decline in their networks as TV aerial reception improved, and who also wanted to steal a march on DBS, which might put them out of business altogether if the price of individual receiving dishes came down. There was also bitter criticism of the decision not to recommend wide-band optical fiber from the influential magazine, *New Scientist*, and from Lord Weinstock, managing director of Britain's GEC. *New Scientist* described the report as "technically inept" and warned that, if accepted, ITAP's plan would "lock Britain into obsolescent technology." Nevertheless, the ITAP report was immediately welcomed by the UK government.

On the same day as the ITAP report was published (March 22, 1982), the Home Secretary announced that Lord Hunt would conduct an inquiry into the "broadcasting aspects" of cable, including standards, methods of regulation, the impact on the BBC and the question of advertising. Already it was clear that there was a difference of opinion between the information technology visionaries of the Department of Trade and Industry and the traditionalists and BBC supporters in the Home Office. Hunt reported swiftly on October 12, 1982, recommending ten-year cable franchises, a central authority to award them, no limit to the number of channels, certain decency standards, advertising, and the compulsory carrying of BBC and ITV channels. In sum, Hunt was advocating minimum restrictions.

The publication of the Hunt Report was followed by criticisms that the government was trying to "rush through" cable TV without adequate controls. Subsequent argument hinged around the possible erosion of broadcasting standards, the use of obsolete technology, the impact of DBS, the huge costs of cabling and possible charges to consumers. One important technical argument concerned the relative merits of the conventional "tree and branch" network and the more sophisticated "switched star" system, which is more sensible for two-way communication and non-broadcast services. But, despite these unresolved questions, there was a mad scramble to form consortia and make bids by both existing operators, BT and numerous others like Thorn EMI and Rank, who hold vast film and video libraries.

By November 1982, serious doubts were being raised about cable TV's potential advertising revenue in the UK and by a CIT Research paper suggesting that the British public wasn't prepared to pay much for the privilege of receiving it. The Department of Trade and Industry decided to duck the issue of which technologies to use, and on December 2, 1982 the government gave the go-ahead to cable TV based on certain minimum standards: 25–30 channels, including DBS, teletext and BBC and ITV; two-way interactive capability; and a preference for "switched star" over "tree and branch," though not of optical fiber over copper coaxial. There would also be a new cable authority and cable companies would not be able to obtain exclusive rights to major sports events.

The arguments raged in Whitehall and beyond, but by April 1983 the government was ready to issue a White Paper which again emphasized the need for quick action, private finance and "light" controls. However, the main proposal was that interim licenses would be issued for 12 pilot systems. That summer a Cable Bill went through Parliament, and the rival consortia (by this time, including showbiz and sports stars, TV newsreaders and former Beatle Ringo Starr) made their bids. In September 1983 the Economist Intelligence Unit was brought in to assess the applications, and in November it was announced that only 11 had met the criteria set by the government.

Within weeks, serious doubts were again being expressed about the economics of the projects, and a new threat emerged in the form of a Treasury proposal to end the 100 percent capital allowances on laying new cables, which would have the effect of extending the break-even point from seven to nine years. That threat was carried out in the Chancellor's spring 1984 budget, and about the same time CIT Research came out with another report suggesting that cable TV in Europe was destined to be a commercial failure under existing policies. One by one, the blows rained down: first the US firm Jerrold pulled out of a joint venture with GEC; then Thorn–EMI and BT announced a "review" of their plans; next, a company formed by Plessey and Scientific Atlanta to supply cable equipment folded. Clyde CableVision (Glasgow) and Merseyside CableVision (Liverpool) had trouble raising money to get started. By October 1984, old-established operator Visionhire announced that it was pulling out of cable altogether, and – to everyone's amazement – even Rediffusion decided to call it quits.

In 1985, TEN, the British cable movie channel, collapsed, Greenwich cable TV cut staff by 50 percent and Thorn–EMI sold out its cable interests. Yet the ITV companies announced grandiose plans to launch a European cable and satellite TV, SuperChannel, and the UK government injected £5 million aid into the cable industry. In early 1986 a report even showed that cable TV viewing figures were slightly up. Cable was still alive – but it was all a far cry from the euphoria of the early 1980s.

Satellite Communications

After digital electronics and fiber optics, communications satellites are the third new technology of the telecommunications explosion. Satellites are pieces of hardware that sit in geosynchronous orbit 22,300 miles above the Equator and thus appear to be stationary when viewed from Earth. Signals – of increasingly higher frequencies – are bounced to and from them via antennae on the ground. Communications satellites are used primarily to transmit long-distance phone calls and to relay radio and TV programmes; but they can also carry facsimile, electronic mail and teleconference transmissions, and they are being used more and more for high-capacity, long-distance data links between computers. Satellites also make possible high-definition TV. At the last count, there were nearly 40 satellites aloft, carrying over 700 transponders. Each transponder has the capacity to transmit one color TV channel, 1500 simultaneous telephone conversations or 50 million bits of data per second.

The original orbiting satellite – though a "dumb" one – was the Soviets' *Sputnik I*, which so astonished the world when it was launched in 1957. The US Army-built *Score* followed in 1958, but it was Bell Labs' *Telstar* in 1962 which began the first regular voice and picture transmissions. A chart-topping record was even named after it. In the same year, Congress formed the Communications Satellite Corporation (Comsat), a private company designed to be the basic US "common carrier" for satellite telecommunications. In 1963 the US-made *Syncom II* became the first satellite to go into geosynchronous orbit, and *Early Bird* (1965) was the first of 12 satellites used by Intelsat, the international communications consortium linking 108 nations, formed in 1964. Intelsat had the international TV and telephone markets pretty much to itself for many years, but in 1985–6 it was engaged in a bitter fight against the Reagan administration's attempts to break it up and introduce more private competition.

Early Bird measured just 28 in by 23 in, weighed 39 kg and could handle a mere 250 phone calls; whereas today's *Intelsat VI* is 39 ft square, weighs 1.7 tonnes and can handle 35,000 calls. By 1972, the Canadians had begun a domestic satellite telephone service via their *Anik* satellite, and in that year the FCC in effect deregulated the satellite industry in the USA by declaring that it would henceforth be developed by private enterprise companies in competition with each other and with other forms of communication.

As demand for satellite services such as rapid data transmission grew in the 1970s and costs came down, the number of satellites in space went up. In 1974–6, Western Union, RCA and Comsat each launched two satellites, and RCA made a hit with its direct satellite transmission, via Home Box Office cable TV, of the Ali–Frazier fight – the "thriller in Manila" in 1975. The Marisat ship-to-shore satellite service got underway in 1976, and in 1978 Eutelsat, representing European PTTs, put up its Orbital Test Satellite which led directly to the European Communications Satellite (*ECS-1*) in 1983 and the current L-Sat advanced satellite technology project.

Satellite Business Systems (SBS), a partnership of IBM, Aetna Life and Comsat, launched its first business communications satellite in November 1980, and in 1981 the Space Shuttle *Columbia* had its maiden flight, opening up a whole new era of satellite launch and recovery. (Tragically, that era came to an end with the shuttle disaster of January 1986.) The number of satellites in orbit is still expected to swell to 40 or more by the end of 1988, carrying a record 900 or more transponders. The more optimistic forecasts put the total of transponders at 1250 by 1990 and at over 3000 by the year 2000.

However, satellite operators have not had it all their own way. The boom in demand for satellite capacity in the late 1970s actually led to the auctioning of transponders at Sotheby Parke Bernet, New York, in November 1980. But by the end of 1982 demand slumped and there was talk of a major shakeout. The price of transponders has since drifted down to $10–11 million, compared with $12–13 million in the late 1970s. The "Big Eight" today in the US satellite industry are SBS, Comsat

(*Comstar*), AT&T (*Telstar*), Western Union (*Westar*), RCA (*Satcom*), Southern Pacific (*Spacenet*), GTE (*G Star*) and Hughes (*Galaxy*).

But the picture is changing all the time: General Instrument Corporation has recently bought into the business through its subsidiary, USCI; ISACOMM rents circuits and resells them, rather like the independents on the ground; and Vitalink and RCA's Cylix offer low-cost data transmission via satellite; in 1984 Comsat ran into cash problems, threatening its plans for direct broadcasting by satellite (DBS), and newcomer Equatorial Communications Co. is doing well with its low-power system which offers a cheaper way of moving masses of data between business districts.

In Europe, development of the satellite industry has been slower than in the USA because of the shorter distances involved, plans for cabling using fiber optics, PTT controls and, most serious of all, national rivalries. The result is that Europe is split into rival French, British and German-led satellite-building consortiums and is struggling to catch up with US companies, especially Hughes, the clear world leader. The European Space Agency – still not fully recovered after a succession of disasters when its *Ariane* rocket crashed into the sea off French Guiana with $ millions worth of satellites on board – is trying to bring the national rivals together. A remote sensing satellite for coastal, ocean and ice observation, *ERS-1*, is its latest joint venture. Meanwhile, Eurosatellite (France, Germany), Unisat (Britain), Telecom (France) and GDL (Luxemburg) have been vying with each other to get their satellite projects operational.

The major current uses of satellites are telecommunications, television, data processing, video conferencing, orbital photography and remote sensing. Telephone traffic is the bread and butter of satellite communications (more than half of all transatlantic calls are now bounced off satellites), but the future very much depends on what happens in fiber optics. Similarly, the prospects for satellite TV or DBS are inextricably bound up with the future of cable TV. Business uses have been slow to take off in some ways (as has DBS itself), mainly because the "office of the future" has been delayed. But satellites are now widely used in the newspaper industry for transmitting copy to remote printing plants (for example, the *Wall Street Journal*, *USA Today* and the *Economist*), while manufacturing companies like Boeing, Westinghouse, GM and GE are using satellites to transmit data such as programs and drawings from data banks located halfway round the world.

Banks, insurance companies and credit card concerns are using satellites for transferring money, calculating policy premiums and "instant" credit-checking. Shipping firms are using satellites for communications and stock control; and even the churches are using satellites to beam their message across continents and oceans. Orbital photography and remote sensing are very promising growth areas: already, devices like the US *Landsat* and French *SPOT* satellites are being used to monitor the weather, ice flow movements and crop growth – a valuable source of information in the Third World. More sophisticated devices are being developed which can spot oil and other mineral deposits from space, while nobody really knows the true extent to which satellites are being used as "spies in the sky"; all we do know is that satellite snooping by the major powers is being carried out on a colossal scale – the Pentagon's Big Bird, for example, is reckoned to be able to read *Pravda* headlines and Zil number-plates in Red Square!

Some have suggested that the scene is set for a titanic struggle between satellites and cable similar to that fought out between the railroads and canals in eighteenth- and nineteenth-century Europe. Optical fibers are challenging satellites, even over long distances, as the best method of relaying phone calls. Cable TV companies are threatened by DBS and the availability of cheaper receiving dishes. But in many instances the alleged rivals complement rather than compete with each other, and for most forms of business communication we are witnessing an *integration* of satellites and fiber optics to create a cost-effective combination particularly well suited for bypassing the local loops in city business districts.

One major regulatory problem looming is what to do with our increasingly overcrowded skies. As satellite slots in the ring above the Equator are rapidly occupied, there are fears that some nations – particularly the emergent nations of the Third World – will not be able to get into orbit. The World Administrative Radio Conference is trying to divide up the ring on an equitable basis. Meanwhile, new technologies may solve the problem before too long.

If higher-frequency bands like Ka-band are used in satellite communications rather than the current C-band and Ku-band, this will enable satellite antennas to form much narrower beams and thus cast much smaller shadows on the Earth (150 miles in diameter rather than continent-sized). This would create more space and also reduce interference. In addition, new coding techniques, better thruster sytems and other advances will serve to increase capacity and performance. The Japanese are already using Ka-band and Europe has L-Sat, due for launch in 1987, which carries an experimental Ka-band package.

Another answer to the satellite traffic jam is to use giant space platforms that will carry hundreds of transponders, thus enabling huge amounts of data to be transmitted to and from the same slot in the equatorial ring. NASA has devised just such a scheme and had plans to launch its Advanced Communications Technology Satellite (ACTS) before the end of the decade. This would be a step in the direction of platforms and would also provide an opportunity to test new materials and technologies like gallium arsenide field effect transistors (FETs). Hughes Galaxy has also announced plans to build a Ka-band satellite.

DBS

Like cable TV, satellite TV or DBS has held great promise, but we are still a long way from the concept of "world TV" programs or the "Global Village" of Marshall McLuhan.

In June 1982, the FCC gave the go-ahead for DBS, designed to serve the 35 percent of the US population living in remote rural areas which are never likely to be reached by cable TV. Eight subsequent applications to provide DBS services were approved, including those of RCA, Western Union and one by Comsat (through Satellite Television Corporation), which proposed using high-power transmitters. Comsat remained in the lead until General Instrument, through its new subsidiary USCI (United Satellite Communications Inc.), stunned the broadcasting world with the announcement in early 1983 that it would begin TV transmissions using medium-power satellites by the end of 1983. In June 1983, the Space Shuttle *Challenger* went up, taking with it the Canadian satellite *Anik-3* which was to be used by USCI, and in November 1983 USCI started America's first DBS service. Customers got five channels of television in return for $300 to cover the cost of the receiving dish and $40 per month rent for a signal unscrambler.

Immediately, Rupert Murdoch's Inter-American Satellite Television Inc. postponed plans to introduce its "Skyband" service and Comsat started to look for partners to help finance its high-cost, high-power system. USCI, meanwhile, was paying the price of being pioneer in the field: its receiving dishes were not easy to install, sales and service were difficult to arrange, and programming was a problem. By March 1984 USCI only had 1000 paying subscribers, and in July 1984 Western Union, CBS and RCA took the hint and effectively pulled out of DBS by failing to file applications on time with the FCC.

In only one respect had DBS really taken off in North America: satellite "space piracy" – the (illegal) interception of satellite transmissions using home TVRO (Television Receive-Only) sets – became something of a cult craze. In April 1986, for example, a "Captain Midnight" broke into a satellite broadcast of Home Box Office to protest about HBO's charges and its decision to "scramble" its signals – thus making satellite broadcasting piracy impossible.

As many as 1.5 million TVRO dishes and thousands of unscramblers have now been sold in the USA alone. Global eavesdropping has become big business, and this has led to allegations of theft and espionage. The Soviets even complained that amateur enthusiasts in the West had been listening in on the back-chat on Soviet TV stations, where announcers had been heard to curse the authorities and recount their sex lives, thinking they were off the air!

Japan launched its first satellite suitable for DBS in early 1984 for the sole use of the Japan Broadcasting Corporation (NHK). The *BS-2a*, as it was called, was designed to improve TV reception for Japan's outlying islands and mountainous regions (see figure 4.3). But it was a disaster: two of its three transponders failed, and the one remaining broadcast channel was not able to reach the widely dispersed highlands and islands of Japan. The fault was traced to some travelling wave tube amplifiers made by the *French* electronics company, Thomson CSF. . . .

In February 1986 the Japanse put up their *BS-2b* satellite, using a Delta rocket made under license from McDonnell Douglas. *BS-2b* was made in the USA by General Electric, but over the next few years Japan intends to go it alone: NEC will build Japan's next broadcasting satellite, and the Japanese National Space Development Agency (NASDA) will develop a new rocket capable of putting satellites weighing 2 tonnes into orbit. A whole series of communications satellites is due for launch by private Japanese companies between now and 1990.

In Britain, satellite TV has been slow to get off the ground, despite initial government enthusiasm. In May 1981 a Home Office report on DBS found support for a project in the aerospace industry – though some skepticism among broadcasters – and urged an early start. In March 1982 the UK government gave the go-ahead for the BBC and Unisat (a consortium formed by GEC-Marconi, British Telecom and British Aerospace) to begin DBS, amid great excitement over this new "license to print money" (to quote the *Sunday Times*). Like cable, there was much talk of the DBS "revolution" and the "new frontier" in space. Satellite-building would also give a great boost to Britain's information technology industry in what had been declared "Information Technology Year" (1982).

But in July 1982 the Home Secretary commissioned former senior civil servant Sir Antony Part to report on the technical standards for DBS, and in November 1982 Part came out for C-MAC, a more expensive and sophisticated system than E-PAL, favoured by the French and Germans. By early 1983 the BBC was having serious second

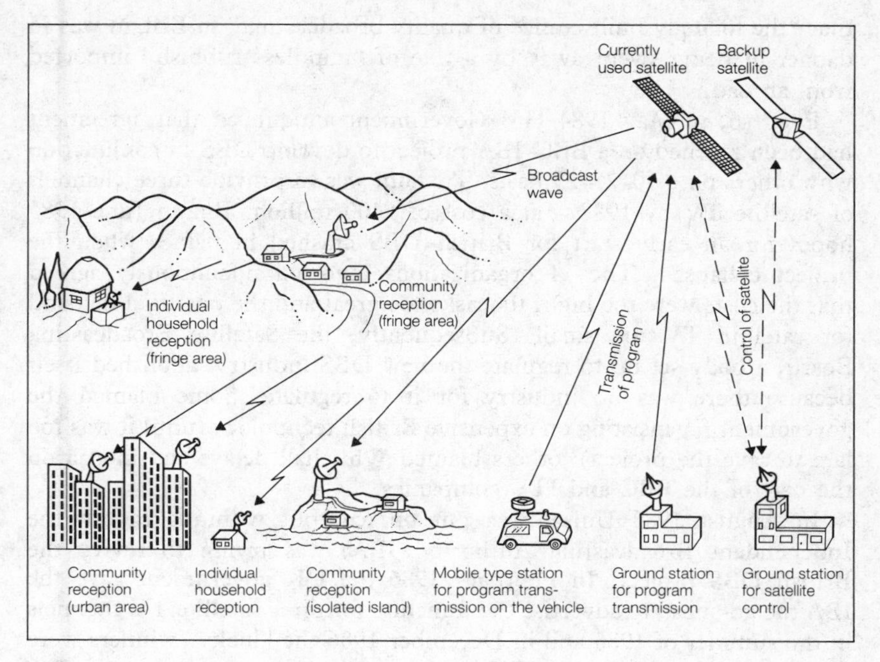

Figure 4.3 Japan's DBS project.
(Source: Report of the Japanese Research Committee of Radio Wave Utilization and Development, 1982)

thoughts about Unisat and in April 1983 it announced an "urgent review" of the prospects for DBS. In August 1983 the BBC said that it could no longer afford even two channels of DBS and asked the government for help in defraying the costs of C-MAC and Unisat, now put at £350 million. The IBA was also approached and was offered two of the five channels of DBS allocated to the UK by international agreement. The IBA indicated that it might make a contribution if the ITV companies could have the renewal of their franchises deferred.

More crisis meetings and technical wrangling followed in late 1983 and early 1984. Some advised waiting for high-definition TV. Others suggested that neither C-MAC nor E-PAL was needed anyway, because satellite TV worked perfectly well on medium-power or quasi-DBS systems. And having viewed Rupert Murdoch's Sky Channel, which began transmissions to Europe in 1982 and Britain in 1984, distinguished broadcasters like Alistair Milne, BBC director-general, joined in a chorus of criticism about the threat posed by DBS to broadcasting standards. An "Open Skies" policy, he warned the Royal Television Society, meant that "the lovingly built edifice of quality broadcasting" in Britain was in danger of being swept away by a tide of "mindless rubbish" imported from abroad.

Even so, in May 1984 HM Government announced that agreement had been reached on a BBC–IBA project to develop DBS in conjunction with others on a 50–25–25 basis. The aim was to provide three channels of satellite TV by 1988 – at a cost of £400 million. But in June 1985 hopes for an early start for British DBS crashed in flames when the project collapsed. The 21 organizations involved unanimously agreed that the costs were too high, the risks too great and the potential demand for satellite TV too small. Subsequently, the Satellite Broadcasting Board, a body set up to regulate the new DBS industry, abolished itself because there was no industry for it to regulate. Some blamed the government for insisting on expensive British technology (until it was too late to save the project); others blamed Whitehall delays and inertia on the part of the BBC and ITV companies.

In August 1985 Unisat was put on ice, but within a month the Independent Broadcasting Authority (IBA) was trying to revive the British DBS project. In February 1986 the UK government gave the IBA the go-ahead to advertise the franchise for three satellite TV channels in the summer of 1986 and in December 1986 the "lucky" winners were selected. Meanwhile, the ITV companies had announced a satellite SuperChannel, which was to combine the best of ITV and BBC programmes and would compete directly with Rupert Murdoch's increasingly successful Sky Channel. The BBC also announced plans for a world TV news service. In a third major development, British publishing magnate Robert Maxwell said his Luxemburg-registered ESTBC (European Satellite Television Broadcasting Corporation) would start a DBS service using a French satellite in the spring of 1987.

In Europe, DBS was initially caught in a similar web of technical tangles and regulatory wrangles with political and cultural overtones. Considerations of domestic industrial strategy also obscured the common interest. Although the Franco-German consortium, Eurosatellite, went ahead with its TDF project, both the French and German governments had second thoughts about taking DBS any further; their main interest for a while shifted to telecommunications.

European-wide advertising has also proved to be a controversial issue: satellite "footprints" are no respecter of national boundaries, and one nation's programs can "spill over" into a neighboring country. For example, a Kellog's Cornflakes commercial shown in Britain could be banned in Holland because it boasts of "extra vitamins," outlawed in West Germany because "the best to you" sounds like a competitive claim, and off the air in France because children aren't allowed to endorse products (in Austria children cannot appear at all – they have to be played by midgets!).

In 1984 Eutelsat (representing European PTTs) became involved in a further argument with Luxemburg's RTL, whose proposed (and partly American-backed) Coronet DBS service threatened to blanket Europe with what was described as "Euro-pap." France's PTT minister Louis Mexandeau thundered that France would not allow Luxemburg's "Coca-Cola satellites" to erode "the artistic and cultural integrity of France." Meanwhile, Murdoch's London-based Sky Channel was slowly and steadily building up its audience in Europe, especially in Holland and Scandinavia.

During 1985 and 1986 there were definite signs of a broadcasting revolution in Europe. Belgium, Holland and Spain opened up commercial television in the space of a few months, while France under President Mitterand underwent a change of heart and announced no less than 85 new private TV stations. Eutelsat and the EEC were both talking DBS again. All were clearly spurred on by the prospect of competition from Sky Channel, Luxemburg's successor to Coronet (called SES) and the Robert Maxwell-led ESTBC, which by 1986 had Italian, German, French and Spanish backing. There was even talk in the UK of the BBC taking commercials.

Cellular Radio

The widespread use of mobile phones and portable phones is revolutionizing communication patterns on a scale almost comparable with the invention of TV, radio or the telephone itself. Providing telephony on the move has suddenly become a multi-million dollar business. As mobile phones become matter-of-fact and cordless communication common, even Dick Tracey's famous wristwatch telephone is a distinct possibility. Soon anyone, anywhere, will be able to call any number anywhere in the world. And it has all been made possible by a remarkable technique called cellular radio, which rates as the fourth new technology of the telecommunications explosion.

Cellular radio works by dividing areas of a city or county into radio transmission zones known as "cells" – hence the name, "cellular" radio. Each cell is served by its own low-power transmitter, which is usually located on top of a tower or high building. Cells are typically two miles across in cities, but may be ten miles wide in rural areas. Each transmitter is only powerful enough to transmit to its assigned area; this means that transmitters in other cells (though not adjacent ones) can use the same frequency in their areas (see figure 4.4(a)). A cluster of cells is controlled by a central computer which routes the calls through the public telephone system. As a vehicle moves from one cell to another, the call in progress is automatically transferred to the transmitter in the neighboring cells as the signal strength weakens (see figure 4.4(b)). This transfer is called a "hand-off" and takes place without the user being aware of what is happening because there is no break in communication.

The idea of frequency re-use employing the cellular technique is an old one, first conceived in 1947 at Bell Labs. But although the concept is simple, its operation is complex, and it was not until advanced electronic switching systems, frequency synthesizers and microprocessors became available in the early 1970s that cheap cellular radio became a reality.

Under the old mobile phone system, as operated by AT&T after 1964, one high-power transmitter covered an entire service area. Because of the limited number of frequencies available and the fact that each mobile phone required the exclusive use of one channel, this meant that in the whole of New York, for example, 700 well-heeled customers had to share just 12 channels, meaning that no more than 12 conversations could take place at the same time. Other areas had as many as 44 channels; but cellular radio – using the old UHF band – has boosted the number channels typically available in a system to 666 – and each one can be re-used. A cellular system can easily handle 50,000 calls an hour, compared with only a few hundred under the old system. It can also be used for data, telex and videotex transmission. The computer control system is the key to the efficient operation of cellular radio.

Although AT&T first proposed a commercial cellular system in 1971, it was not until 1974 that the FCC responded and 1977 before the FCC issued a license to AT&T to start a trial system in Chicago. This actually began in 1978 and was followed by further experiments in the Washington–Baltimore area and in Raleigh–Durham, North Carolina. The Chicago trial was an instant success and quickly attracted 2000 customers. In 1982 the FCC finally got round to resolving the legal and regulatory wrangling between the phone companies or "wireline operators" and the "non-wire" companies or radio common carriers (the RCCs). It decided to grant *two* licenses in each of the first 30 cities to have cellular radio: one would go to the existing local phone company, the other to one of the (competing) RCCs.

So keen was the competition to start cellular radio services that the FCC was swamped with multiple license applications when it opened for business on June 7, 1982. Up front in this "Gold Rush" were the regional phone companies-to-be, like Ameritech and Atlantic Bell,

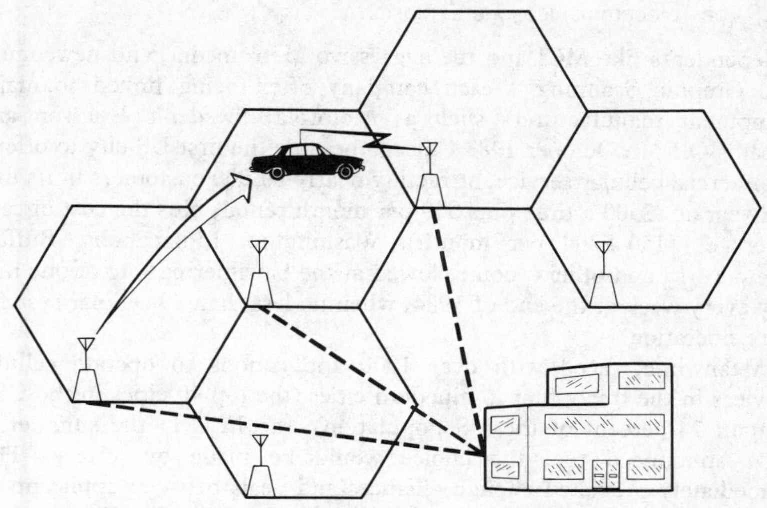

(a) Advanced mobile phone service. As vehicles travel from one cell to another, electronic switching equipment automatically hands off calls, without interruption, from one radio transmitter to another.

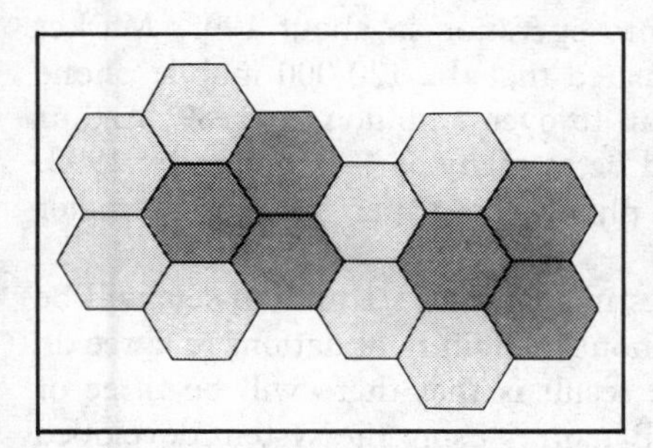
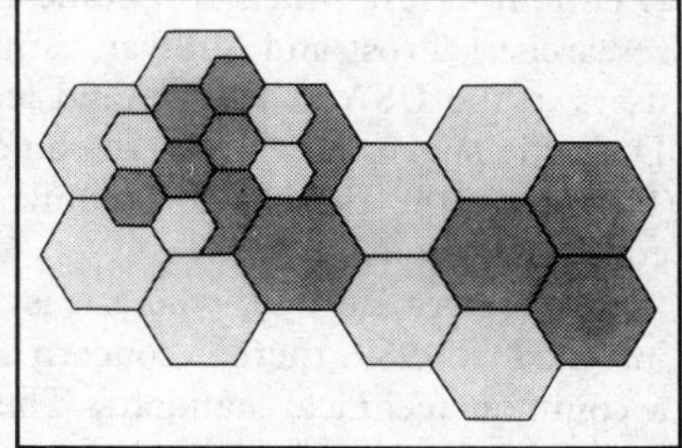

(b) The advantage of cellular radio for mobile telephone systems is "frequency reuse": a limited spectrum of radio frequencies can be made to serve many users. The area to be served is divided into cells – in this case, 14. A low-power radio transmitter – powerful enough to reach any mobile unit in the cell but not powerful enough to interfere with transmissions in distant cells – is located in each cell. Cell-splitting is a simple way to keep up with growing demand for mobile telephone services in a cellular system. The right-hand diagram shows how nine cells have been created out of parts of four cells.

Figure 4.4 How cellular phones work.
(Sources: (a) "Evolution of the Intelligent Communications Network" by J. S. Mayo, *Science*, vol. 215, February 12, 1982, pp. 831–7. Copyright, 1982 Bell Telephone Laboratories Inc., reprinted by permission. (b) *Technology Review*, November/December 1983. Reprinted by permission of *Technology Review*. Copyright 1983.)

independents like MCI and the aggressive Metromedia, and newcomers like Graphic Scanning – each company often being linked to major equipment manufacturers such as Motorola, Sweden's Ericsson and Japan's Oki. In October 1983 Chicago became the first US city to offer a commercial cellular service, attracting nearly 10,000 customers in its first full year at $2500 a time plus $50 per month rental, plus the cost of calls (average: $150–$200 per month). Washington, Indianapolis, Buffalo (New York) and others soon followed at the bewildering rate of one new city every week at the end of 1984, when no less than 30 cellular systems were operating.

Meanwhile, faced with over 1000 applications to operate cellular services in the thirty-first to ninetieth cities (the top 90 cities in the USA contain 74 percent of the US population), the FCC in the summer of 1984 announced that the choice would be made by lottery. This immediately provoked enough alliances and deals between applicants to obviate the need for a lottery – which suited the applicants especially well, because they were eager to get started before or at the same time as the local phone companies. The FCC then turned its attention to the 5200 applications for licenses in the ninety-first to one-hundred-and-twentieth cities, which will come into operation in about 1995. Market researchers Frost and Sullivan estimated that the 120,000 mobile phone users in the USA in 1984 would leap to over 2 million by 1989; Arthur D. Little suggested that there could be as many as 4–6 million by 1992. That puts the market for mobile phones on a par with other major consumer goods.

In Western Europe, where it is estimated that 1.1 million sets will be installed by 1989, there is concern about the failure of nations to agree on a common technical standard. The result is that there will be three or more incompatible systems in use. Britain is using the system developed in the USA by AT&T and Motorola called AMPS and renamed TACS (Total Access Communications System) for the UK market; the Scandinavians are using their NMT (Nordic Mobile Telephone) system; the West Germans initially went for the Siemens NET-C system and then in early 1984 joined with the French to develop a version of the Philips/CIT–Alcatel MATS-E system, previously rejected by the French government.

The Scandinavians were first in the field with the L. M. Ericsson NMT system, which went commercial in 1982 and had over 100,000 subscribers in Finland, Norway, Sweden and Denmark by 1984 – making it, in fact, the largest and fastest-growing cellular system in the world. A modified version of NMT, using the 900 rather than the 450 MHz frequency, is now being phased in. NMT or modified NMT has been adopted by Austria, Holland, Ireland and Spain, with Italy, Switzerland and Greece yet to decide. Under the Franco-German agreement, four pilot schemes are being run in Paris, Lyon, Frankfurt and the Ruhr before cellular goes nationwide in France and Germany, but only about 100,000 subscribers are envisaged by 1989 in these two nations.

British Telecom submitted a plan for cellular radio in 1980, but it was not until June 1982 that the British government announced plans to license two competing consortia to run mobile phone services: one would be BT in association with the security firm Securicor, and the other would be chosen later. In September 1982 the Merriman Committee gave strong support to the development of cellular radio on economic grounds and recommended in its report the use of the soon-to-be-vacated 405-line TV frequency bands. By December 1982 the government was able to give the go-ahead to the BT–Securicor and (to everyone's surprise) a Racal–Millicom consortium to operate cellular services.

But first, the British government had to decide whether to use the unproven but European MATS-E system or the Americn AMPS system, renamed TACS. In February 1983 it plumped for TACS on the grounds that it made possible an early start for cellular in Britain – set for January 1985. Racal–Millicom (now called "Racal Vodafone") meanwhile lined up Sweden's L. M. Ericsson as switching equipment supplier and ordered phones from Japan's National Panasonic and Finland's Mobira. TSCR (Telecom Securicor Cellular Radio) said it would use mainly Motorola equipment and NEC phones for its "Cellnet" service.

Both consortia rushed to get started in January 1985, and both spread out rapidly from London, attracting a total of 45,000 customers by the end of the year. This unexpected success for cellular led to predictions that as many as 500,000 cellular car phones would be installed by 1990. Cellnet had 58 percent of the market in 1985, but as its share dropped in early 1986 it started a price war which led to a shake-out among mobile phone retailers. Also, both Cellnet and Vodafone increased phone charges.

The Japanese were quick to spot the potential of cellular radio, which had obvious attractions in a nation of densely populated urban areas. NTT set up its first radio telephone service in 1979. By 1984 it had an extensive cellular network , with 300 systems serving all of Japan's major cities. Even so, only 27,000 customers had signed up, although 10,000 new subscribers were expected to join every year. Privatization of NTT was expected to give the market a boost, but meanwhile Japanese manufacturers, as ever, looked to overseas.

The opening-up of the mass market for mobile phones means that they are no longer a rich man's toy. They are becoming an essential working tool for executives, salespeople, fleet operators, plumbers and professionals like surveyors. The vast transport and distribution and construction industries are now major users, and most customers are citing considerable gains in efficiency and productivity as a result of using cellular radio. In Chicago, for example, a fast food chain reported savings of 1 million gallons of gas a year; a trucking company increased its business by 30 percent; real estate executives and travelling sales people reported 20, 30 and 50 percent gains in productivity as they extended their working day in the car by not, for example, having to return to the office to make calls.

Like most new technologies, cellular radio is not standing still. The concept is being developed to include truly portable, battery-powered, lightweight pocket phones at reduced cost, the facility to "roam" from one area to another using the same phone, and an all-digital "second generation" of mobile phones which will increase the number of calls possible in a cell by reducing the amount of radio spectrum required. British researchers are working on plans to transmit via satellite, and there are mobile systems planned for factory use – linking, for example, fork lift trucks to the warehouse computer. In addition, cells can of course be split and split again to increase capacity, and a new radio version of packet-switching will make possible the use of cellular radio for rapid data transmission to mobile terminals.

But there *are* alternatives to cellular. In the USA there are current proposals to start PRCS (Personal Radio Communications Service), which is designed to provide a cheaper way of linking people within a limited range of their home or office. Cellular radio provides a high-quality service, but it requires an elaborate system of transmitters, and the phones are still quite expensive to buy. PRCS systems rely on a small base station installed in the user's home or office and will be cheaper all-round. They can also be used as a telephone or as car-to-car radio.

Britain has raced to develop non-cellular alternatives. In May 1984 the UK government issued a consultative document inviting proposals for the wider use of mobile radio using the Band III radio frequency made available by the ending of 405-line TV transmissions. Despite delays in 1985, the go-ahead was given in March 1986 for two national private mobile radio systems to be run by Pye and GEC. In addition, a number of local systems were sanctioned. Consultants CIT Research were predicting 750,000 private mobile radio units in the UK by 1990. Band III devices are not connected to the telephone network: they are used by large organizations or transport and distribution companies, enabling their vehicles to contact base.

Meanwhile, there has been great excitement over the introduction in the USA of Airfone, which enables passengers on 13 airlines to make and receive calls from 35,000 ft up. The system operates through a network of 49 ground stations spread across the USA which can transmit and receive calls at a distance of 400 miles. Airfone initially ran into trouble with the FCC, which said it was wasting scarce frequencies, but the Commission later relented. In the UK, British Airways (BA) hopes to follow suit in 1987 with an international in-flight phone service using an Inmarsat satellite.

New Network Services: Electronic Mail, Videoconferencing

The new technologies of the telecommunications revolution have made possible the provision of new digital communications services – sometimes called Valued Added Network Services, or VANS – which use the basic telephone network but provide something that is not normally obtainable from it. They therefore "add value" to information *en route*. Major categories include electronic mail, videoconferencing and videotex, which are examined here; commercial reservation systems, remote EFT and credit card transaction systems, databases and financial information services are looked at in later chapters.

Of course, computers have communicated over the telephone network for years, but the new network services enable third parties to buy, sell or exchange goods, services or information without having to rely on the mail, telephones or face-to-face contact. For example, AT&T's Tradanet enables shops to order goods directly from manufacturers on the network, based on information from laser scanners reading the bar codes on goods passing through the checkouts.

Network services have great potential, but until recently their development in most areas was disappointingly slow. In the UK, for example, progress has been delayed by government indecision over regulation. Now things are looking up as transmission techniques improve and computer costs come down. But a looming clash of incompatible communications standards will first have to be resolved: AT&T and others are adopting the Open System Interconnection (OSI) model, which will be a set of standards governing everything from the number of pins on a plug to the way in which programs work. IBM, meanwhile, has adopted its own communications standard called Systems Network Architecture (SNA), which it is encouraging others to use so that they will have to buy IBM equipment – although there are now signs that "Big Blue" will go over to OSI. The coming of the Integrated Services Digital Network (ISDN) in the next decade is expected to give a major boost to the provision of network services of all kinds.

Electronic Mail

The term "electronic mail" in its broadest sense covers all types of electronic messaging, including telex, facsimile and mailgrams. But, increasingly over recent years, "electronic mail" has come to be associated with so-called "computer mailbox" services, or computer-based messaging systems (CBMS), which enable messages to be transmitted directly between users' computers. There are three major kinds: time-sharing services, run by independent operators; complete software systems on sale to companies; and hybrid systems, also run by independents, which mix electronic inputs with paper delivery. In time, electronic mail systems will evolve to a point where many employees will have a desktop terminal providing access to extensive networks.

Electronic mail at last looks like taking off. For 20 years, it seemed the obvious answer to mailroom and telex room delays, time wasted on unproductive phone calls, and the high costs in executive and secretarial time of dictating, typing and mailing memos and letters. Recent surveys have found that only one in three business calls actually connect to the right person, and the average business letter costs about $10 to produce and transmit. But high installation costs, executive resistance to change and the familiar "chicken and egg" problem of not enough users held things back.

Now the proliferation of personal computers in offices and the heavy promotion in the USA of new services is boosting the number of electronic mail users. MCI Mail, a hybrid system introduced in late 1983, claimed 200,000 subscribers by early 1986. MCI did face stiff competition from Federal Express's ZapMail (see figure 4.5), which offered combined facsimile transmission with hand pickup and hand delivery (until it was peremptorily closed down in September 1986). In addition, Western Union is spending large sums in developing Easylink, its sophisticated, high-speed communications network which had 125,000 users by early 1986 and has a British link-up through Cable & Wireless Ltd. But all three have run up large losses, and they face further competition from ITT's Dialcom service, AT&T Mail (started in February 1986), and from companies like Tymnet and GTE Telenet, which are also entering the field.

None of them, however, has to worry about competition from the US Postal Service. The USPS started its own service called E-COM (for Electronic Computer Originated Mail) in January 1982, but by April 1982 only 96,000 messages had been transmitted, against a predicted 20 million. Volume eventually picked up to a third of target, but it was later decimated after first the Department of Justice and then the Postal Rate

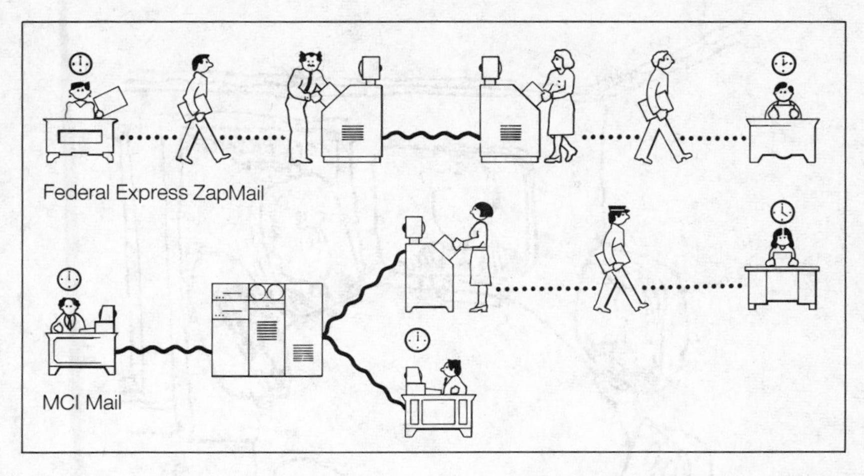

Figure 4.5 Two US electronic mail systems. ZapMail makes minimal demands on sender and recipient. A courier picks up the sender's document and takes it to a Federal Express office, where it is faxed to an office near the recipient. A courier at that end hand-delivers it. Federal Express promises delivery within two hours – and the clock starts ticking when the sender calls. With MCI Mail, a more complex system, the sender keys the message into an electronic machine, and MCI's network can then transmit it instantaneously to the recipient's machine. Alternatively, MCI offers to retransmit it to a processing center for delivery through the US mail or by Purolator Courier. Hand delivery at its fastest occurs within four hours after transmission.
(Source: *Fortune* magazine, August 20, 1984, p. 150. Illustration, Daniel Pelavin.)

Commission forced the rates up, following complaints from competitors that E-COM's rates had been pegged unrealistically low. After running up losses of $50 million, E-COM was discontinued in June 1984.

In Britain, the electronic mail business is booming after a delayed start. The Post Office began its own Intelpost facsimile service in 1980 (which includes international links via Comsat), but private companies were not allowed to sell services using the BT network until 1981. Now five services are on offer: BT's Telecom Gold (based on ITT's Dialcom system, which had 29,000 subscribers in 1983); Istel's Comet service; Geisco's Quik-Com; Western Union and Cable & Wireless's Easylink; ADP's Automail; and Kensington Datacom's One to One (now owned by Pacific Telesis). Although less than 60,000 subscribers had been signed up by 1986, the market is expected to grow 15 times by 1990 as prices come down and the savings become more apparent.

Videoconferencing

It is well over 20 years since AT&T first demonstrated its Picturephone in April 1964 at the New York World's Fair. But as someone once said, videoconferencing (or teleconferencing, as it was called earlier) has "captured more imaginations than it has business" in all that time. The potential for savings in business travel has been obvious: as *Fortune* magazine once put it, "The business of American business is going to endless meetings – what-are-we-going-to-do meetings, meet-the-other-guy meetings . . . even why-are-we-having-so-many-meetings-meetings." Yet 71 percent of executives in one major survey felt that most meetings were a complete waste of time. The general idea behind vieoconferencing is that companies can cut out many of these time-consuming and costly face-to-face meetings and save money; and, with the costs of transmitting images coming down and the costs of air and surface travel relentlessly rising, videoconferencing will come into its own.

That, at least, was the theory. But companies have found the practice of videoconferencing a great deal more complicated and expensive than the tele-visionaries implied. Technical glitches, poor picture quality and psychological problems of relating to other people through a TV screen have proved that there is no real substitute for pressing the flesh. Moreover, executive employees seem to *like* their "business" travel and have resisted trying something new. In Britain, for example, the Post Office launched its Confravision service with a great fanfare in the early 1970s, but it never caught on.

High costs have invariably been the main stumbling block, but now prices are falling rapidly thanks to the reduced costs of satellite transmission and the invention of some remarkable new devices called "Codecs" (for digital CODing and DECoding equipment). In the past, sending a full-motion color TV picture required an entire satellite transponder, representing a huge amount of radio frequency space or "bandwidth." What firms like Compression Labs, GEC Video Systems and Japan's NEC have done with their Codecs is to code the TV signal into digital computer bits and then compress it by removing 99 percent of the information – such as the background, which remains the same for frame after frame. A full motion color TV picture can typically be transmitted via Codecs using only 768,000 bits of data per second (equivalent to just 12 ordinary phone calls), instead of the millions of bits required before. Low-speed Codecs, using only 56,000 bits per second, are even cheaper to run.

As a result, carriers such as AT&T, American Satellite and Vitalink have been able to slash prices, and a number of companies such as Sears Roebuck and J. C. Penney have announced plans to set up or expand their own video networks in the next few years. Retailers find videoconferencing especially useful for displaying new merchandise to buyers and managers, while manufacturing firms like Boeing in Seattle are also getting hooked on video. In the UK, the Royal Bank of Scotland has a very sophisticated system linking its London and Edinburgh offices.

Hilton Hotels have plans to provide public videoconferencing facilities in 35 major hotels. Comsat and Inter-Continental Hotels, a subsidiary of Grand Metropolitan, started a service between London and New York, called Intelmet, in 1983 – but it closed in 1985. AT&T have had better luck with their videoconference service linking business customers in London, New York and 13 other US cities, started in 1984. But in 1985 AT&T had to scale down its public videoconferencing network, Picturephone Meeting Service, linking 11 US cities, owing to lack of demand. When launched in 1982, the company said it would expand the service to 42 cities.

Nobody was expecting videoconferencing to be an overnight sensation, and it faces stiff competition from private satellite networks (PSNs), which are being tried out on an experimental basis. It is therefore still a little early to claim that, because of videoconferencing, business travel will soon be a thing of the past.

Videotex

Videotex has been described by its fans as the spearhead of the information technology revolution that will transform the living room TV set into some kind of electronic supermarket. Critics have called it "a solution looking for a problem" – a technologically glamorous product, but one that is not worth investing in because of its limited uses. Videotex – the generic name for screen-based information systems – never achieved the immediate and widespread success that was predicted for it in the early 1980s. Consumers showed a marked reluctance to pay for its services. But now it is steadily gaining acceptance, not only in the home but more especially among certain kinds of business users.

There are broadly two kinds of videotex. First, there is *broadcast teletext*, in which information is transmitted, using spare lines on a conventional TV signal, to appear on the domestic TV screen. This one-way service – technically described as "receive-only page capture" – is best exemplified by the BBC's Ceefax and ITV's Oracle services, which were started in Britain in 1976. (Teletext is not to be confused with teletex, which is a high-speed text communications link between computers over the telephone network – a kind of super-telex. This has not proved very popular, and in Britain, for example, a government-backed scheme to demonstrate its uses, called Project Hermes, was abandoned in December 1984.)

The second, and better-known, form of videotex is *viewdata*, which represents the marrying together of a TV screen as the display device and the public telephone network as a means of transmitting information from a central computer. Viewdata offers a two-way interaction service suitable for home banking and teleshopping, apart from home information systems, and gives access to information in a variety of computers using "gateway" software. But business uses such as travel reservations, stock location and financial information services are proving more popular than domestic services, and account for a large and growing proportion of videotex terminals. Examples include British Telecom's Prestel service, launched at the end of 1979, Knight-Ridder's Viewtron experiment in Florida (which folded in 1986) and Chemical Bank's

Pronto home banking system, available in the New York City area.

At the last count, there were about 5 million videotex terminals in use in the world: about 1.8 million of these were homes in the UK with teletext TVs and 1.5 million were French Minitel viewdata terminals. About a quarter of US users are commercial databases such as CompuServe, The Source and Dow Jones News Retrieval. There are about 55,000 users of Prestel in the UK. West Germany's Bildschirmtext (BTX) had just less than 20,000 users by the end of 1985.

In stark contrast to the rest of the world the French are riding a veritable videotex craze which they call "Le Phénomène Minitel." It all began in 1981, when the French PTT started giving away small, unsophisticated terminals for use as electronic telephone directories. Soon private entrepreneurs saw the potential and started offering videotex services ranging from news, games, stock prices and credit-checking to medical advice, dating and wife-swapping. Many retailers find Minitel invaluable for taking orders, and French consumers love to order their oysters, foie gras, cognac and Bordeaux direct from the chateau. By 1990 the French PTT expects to have a staggering 10 million households equipped with Minitels.

Despite the slow progress in the rest of the world, some forecasters continue to predict a bright future for videotex. In the USA, Booz Allen & Hamilton Inc. recently reported strong demand from customers and suggested that between 17 and 30 million households would have home information systems by 1993. In the UK, Butler Cox consultants predicted a big market for videotex, although they argued that the "videotex industry" as such would converge with the personal computer and office products industries. Videotex would take its place alongside a number of other communication and display protocols. Diebold foresaw 2.8 million BTX users in West Germany by 1991.

On the other hand, more skeptical analysts, such as CSP International and in particular the Yankee Group, have consistently poured cold water on the predictions of the videotex boosters, highlighting the problems of using videotex – such as its slowness and inflexibility – and pointing out the apparent unwillingness of consumers to pay for mere information. Another obvious problem is the differing standards adopted by nations: the UK has Prestel, France has Teletel, based on the Antiope system, while AT&T in the USA is using a version of Canada's Telidon called NAPLPS (for North American Presentation Level Protocol Syntax).

Prestel, which went commercial in the UK in March 1980, was the world's first public home information system – a pioneer of the so-called "information" industry. Launched with great fanfare, the British Post Office confidently predicted 100,000 users after the first year and a great future for this "world-beater," the development of which had cost the British taxpayer £20 million. But at £600 for the TV adaptor and high telephone charges afterwards, Prestel had few takers, and it turned out to be an expensive flop. By June 1982 only 16,000 customers had signed up, and even today there are only about 51,000 Prestel terminals in use.

The Prestel version of viewdata proved to be slow, expensive and lacking in excitement. Its graphics were crude, and the "tree" system of accessing information was cumbersome and irritating. Consumers mostly found this kind of "infotainment" far from entertaining! During 1981 there was much agonizing about the Prestel flop: the service was reorganized, inquiries were held, and there were cutbacks at Prestel HQ.

In 1982 the first of a trickle of new services – hoped-for saviors of Prestel – started to appear. The Nottingham Building Society, linked with the Bank of Scotland, announced its Homelink home banking service, which offered free Prestel adaptors to potential customers. In 1983 the Birmingham Post and Mail Group, through its Viewtel subsidiary, began the Club 403 teleshopping experiment, which continued after its test run ended in December 1984. Also in 1983, Prestel tried again with Micronet 800, a telesoftware service for personal computer users run by the East Midlands Allied Press. None of these were runaway successes, but they did attract some users and helped to compensate for the withdrawal of some of Prestel's original "information providers," such as the newsagents W. H. Smith, who pulled out in 1984.

There were other modest successes, too. Farmlink, a service to farmers in the West Country, proved popular. And National Homes Network, a service to real estate agents, was working well. Stock market investors were turning to CitiService, and both the hotel and airline industries were making good use of Prestel. Private viewdata systems in the UK – some of whom go through Prestel – also reported growth. Even so, the level of usage was nothing like that originally envisaged, and the critics continued to portray Prestel as some kind of Concorde-style boondoggle.

In the USA, apart from Warner-Amex's Qube cable TV experiment in Columbus, Ohio, which has a home shopping capability, the first real home information system based on viewdata was Knight-Ridder's Viewtron service in south Florida, launched in October 1983. Many saw this as the pioneer of electronic publishing in North America. Viewtron subscribers had to buy a special AT&T terminal called Sceptre for $600 and pay $12 per month, plus telephone charges. In return, they could shop, bank, catch up with the news and look up all kinds of information in commercial data banks. But Knight-Ridder signed up only 5000 customers in the first year, and the cost of providing the service far exceeded the income from users. It folded in March 1986, after Knight-Ridder racked up losses of $50 million.

Times Mirror started its similar Gateway service in 1984, also using Sceptre terminals, in Orange County, California – but it too folded, in March 1986. Centel and Honeywell launched their Keyfax service in Chicago – only to see it fold in 1985. In addition, IBM, Sears and CBS announced a joint venture to develop a nationwide commercial videotex service, called Trintex, for households possessing a home or personal computer – as did CNR (owned by Citicorp, Nynex and RCA) and Covidea (AT&T, Time, Chemical Bank and BankAmerica). The reasoning behind their decisions was that a huge potential audience for videotex was being created by rising sales of personal computers, and the market would take off once the appropriate modems were available. But things do not seem to have worked out that way: CBS pulled out of Trintex and Covidea began laying off workers in December 1986.

Home banking via videotex has been fairly slow to take off in North America and in Europe. The push for banking from home has come mainly from the banks themselves, who see an opportunity to save on staff and paper. But despite some obvious merits, consumers aren't exactly besieging the banks demanding it. Enthusiasts say that, once it is possible to play with your accounts in the living room on a wet Sunday afternoon, it won't be long before people refuse to make the tiresome journey to the downtown or branch bank. Although most banks have explored the idea – and some in great detail – only a few are actually running home banking experiments.

The two most successful such experiments in the USA are Bank of America's service in San Francisco, which boasts 15,000 subscribers, and Chemical Bank's Pronto in New York City, which has attracted 21,000 users since it was launched in September 1982. All the home information systems described above also incorporate home banking functions. Homelink has done modest business in the UK, while the Verbraucher Bank in Hamburg, West Germany, claims 50,000 regular users of its home banking service, which must make it the most popular in the world.

Teleshopping has potential, and has been actively investigated by many companies, especially mail order firms. From Qube onwards most of the videotex services available have had a home shopping facility, though usage has been fairly limited. Comp-u-Store, reached on several US networks, is one service specifically devoted to selling a wide range of goods, and there have been local experiments such as Grocery Express in San Francisco. Club 403 in Britain found that the housebound and the disabled were particularly grateful for the facility.

Business uses seem to be the major growth area in videotex. As we shall see in the next chapter, and especially in chapter 8, databases, financial information services and other forms of electronic publishing are growing rapidly in popularity. Librarians, scientists, doctors and lawyers are increasingly turning to the flickering screen rather than their dusty tomes for up-to-date information. Many analysts are predicting that "electronic publishing" will soon become very big business indeed. The other growth area is in private videotex systems, which are being widely used in the travel business, retailing and in the manufacturing and distribution industries. UK consultants Butler Cox see the number of private systems in operation in Europe growing from only 700 in 1983 to over 15,000 by 1993.

Videotex is probably losing its distinct identity. In the home, it is converging with home computers. In the office, it is merging with conventional office automation systems. From being seen initially as a single public service, videotex is now being used in a great variety of private systems. Its future is thus intimately tied up with the future of information technology as a whole.

5 Personal Computing

Personal Computers – The Personal Computer Boom – The Software Revolution – Taking Computers Home – Consumer Electronics: the Video Age – Working From Home: the Electronic Cottage – Computers in the Classroom

In early 1975, the electronics hobbyist's magazine, *Popular Electronics*, featured a strange rectangular box on its cover, which looked rather like a stereo-amplifier. It was called the Altair microcomputer, and a kit of parts could be ordered for $397 from an Albuquerque firm, Micro Instrumentation and Telemetry Systems (MITS). In made-up form, it cost $621. Based on the first 8-bit microprocessor unveiled by Intel in 1972, the Altair had no keyboard or screen and it wasn't exactly "user-friendly." Even so, it was popular with engineers, scientists and hobbyists, and MITS had sold a respectable 2000 machines by the end of 1975.

Among the electronics enthusiasts attracted to the Altair were a small group of youngsters in Silicon Valley who called themselves the Homebrew Computer Club. Homebrew members first met in a garage in March 1975, in order to discuss and tinker with the latest gadgetry. Excited by the Altair and the idea of "personal" computers, the more ambitious wanted to build their own.

Homebrew included some pretty bright kids – among their number, the founders of IMSAI, Cromenco and Processor Technology; Bill Gates, the boy genius who started software giant Microsoft at the age of 16 and had IBM beating a path to his door by the time he was 24; and, most famous of all, those quirky wizards Steven P. Jobs and Stephen G. Wozniak, who went on to found Apple Computer, the sensational success story and living legend that literally grew from a two-guys-in-a-garage operation to a $1 billion international corporation in just eight years.

The Apple story began, like Homebrew, in a garage – Steve Jobs's garage in Los Altos, California, to be exact. Jobs and Wozniak ("Woz") couldn't afford an Altair, so they set out to build one. Woz was the technical wizard or computer "nerd," while Jobs was the hustling entrepreneur. Using parts from Hewlett–Packard and Atari – where Woz and Jobs worked – and a cheap microprocessor, they first showed Apple I at the Homebrew Club in early 1976. It was an instant hit, and everybody wanted one. By selling Jobs's Volkswagen and Woz's calculator, the dynamic duo raised $1300 and began production. In June 1976 The Byte Shop, an early Silicon Valley computer shop, ordered 50 Apples. They were in business – although still in the garage.

Excited by the prospect of retailing personal computers, Jobs approached his boss at Atari, Nolan Bushnell, and Wozniak his seniors at Hewlett–Packard. They didn't want to know. Then, through local venture capitalist, Don Valentine, Jobs and Woz were put in touch with A. C. "Mike" Markkula, a former marketing boss at Intel. Markkula joined forces as a third equal partner and set about raising more money, hiring key personnel and establishing a manufacturing facility. It was at about this time that the famous Apple logo (a rainbow-colored apple with a bite out of it) was created, with the advice of Valley PR whiz, Regis McKenna.

In April 1977 Apple first showed the Apple II at a San Francisco computer fair. Subsequent sales went through the roof, and by the end of 1977 Apple had sold $2.5 million worth of computers. Not only was it a great technical success, it became the machine that created a mass market for personal computers. The success of the Apple II encouraged thousands of software writers to create programs specifically for use on the Apple, using BASIC, the simple programming language developed by Bill Gates and Paul Allen. This had the effect of boosting sales at a crucial time. Today, there are over 16,000 programs available for Apple computers.

In 1978 Apple's sales hit $15 million; in 1979 $70 million; in 1980, $117 million; in 1981, $335 million and in 1982, $583 million, putting Apple into the *Fortune* 500 of top companies in just five years. By 1983 Apple had nearly 4000 employees and about 750,000 Apple IIs had been sold. A hundred or so Apple employees had also become millionaires when Apple went "public," and Jobs himself – just 28 years old in 1982 – was worth about $300 million, which made him the youngest of the 400 richest people in the USA that year; in February 1982 he even made the cover of *Time* magazine.

Some have suggested that Wozniak and Jobs are geniuses, but Michael Moritz, in his book, *The Little Kingdom: The Private Story of Apple Computer* (William Morrow, New York, 1984), says in effect that they were simply lucky to be in the right place at the right time. However, their complementary talents created an explosive "mix," and it is difficult to imagine Apple succeeding without Jobs's driving ambition. Whatever its founders are or were, the Apple story is a fine example of what could happen a decade or so ago in Silicon Valley. The rise of Apple is also central to the wider story of the remarkable rise of the personal computer.

Personal Computers

In the comparatively short space of a few years, the personal computer has been transformed from a hobbyist's toy to an all-pervasive working tool. Making personal computers, once the prerogative of amateur tinkerers, is now a multi-million dollar industry. In fact, it qualifies as the fastest-growing industry ever. Back in the 1960s, computers could be afforded only by large organizations; then, in the 1970s, departments or sections of organizations were able to buy them; in the early 1980s, personal computers came to be owned and operated by individuals, making a reality of distributed processing.

In offices, personal computers have started a desktop revolution by putting 1950s mainframe computing power on the desks of millions. In the home, they are the biggest thing since the TV and have had an even faster penetration rate than the telephone. They have even been described as the basic building blocks of a new kind of society.

A personal computer is essentially a small computer based on a tiny microprocessor (see figure 5.1). This acts as the central processing unit of the computer, supervising the operations of the entire system. Other chips provide memory for the storage of instructions and data, while external storage devices such as cassette decks and floppy disk drives provide further memory capacity which can be transferred from one personal computer to another. Information is entered into the computer by means of a typewriter-like keyboard or is transferred from disks or tapes. Output is displayed on a TV screen or the computer's own monitor. It can also be printed out on paper or converted from digital to analog signals by a modem for transmission over telephone lines.

These elements constitute a personal computer's *hardware*. But the hardware can do nothing by itself; in order to be useful, the personal computer requires application programs or *software*, sets of instructions that tell a computer what to do. Software packages like Visicalc made personal computers worth having in the first place and have played a key role in stimulating demand for certain models. The core of the software is its "operating system," which controls the computer's operations and manages the flow of information.

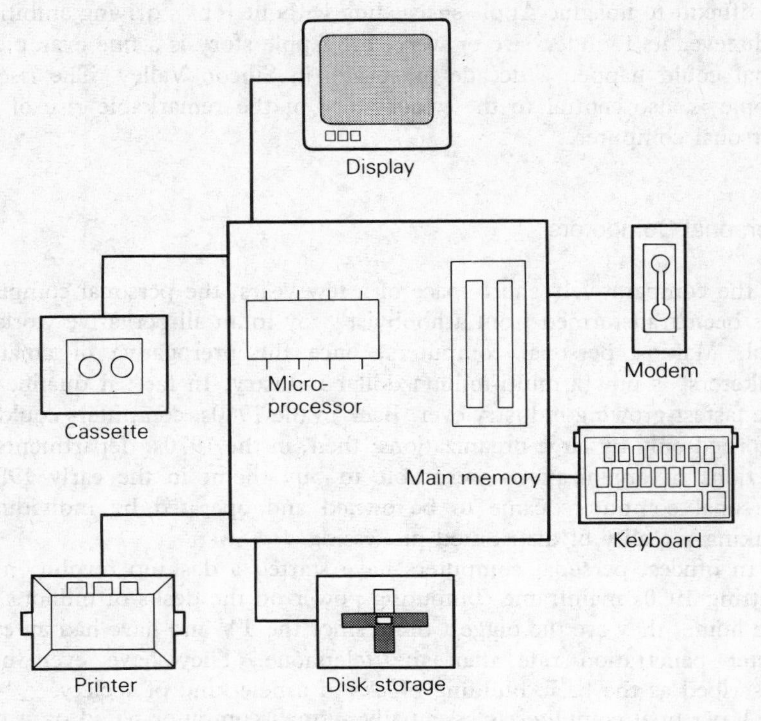

Figure 5.1 A typical personal computer configuration.
(Reprinted with permission, *Technology Review*/Poonam Gupta, copyright January 1983.)

In fact, the most significant factor in defining a personal computer is the characteristics of its operating system. The trend is for personal computers to operate at higher speeds and to use longer and longer "words." The first personal computers used 8-bit microprocessors – with the capacity to process information in "words" of eight digits – while today's models use 16-bit or 32-bit microprocessors. Operating systems are becoming standardized around certain types like Unix (first developed by AT&T's Bell Labs), MS-DOS (IBM's system designed by Microsoft) and MSX (the Japanese standard for home computers, also developed by Microsoft).

Sales of personal computers rocketed from none in 1975 to 7 million units sold worldwide in 1983. In the USA, sales of 47,000 units in 1977 grew to 500,000 in 1980, 750,000 in 1981 and 1.4 million in 1983. The worldwide installed base of personal computers has grown from 200,000 in 1977 to over 50 million in 1987. Over the decade 1977–87, the world market for personal computers has increased in value from a few million dollars to $27 billion, equal to the world market for mainframes and overtaking that of minicomputers (see figure 5.2). By 1990, less than a decade after the first volume shipments, personal computers will account for nearly 40 percent of all computer shipments by value and a third of the value of the entire installed computing base (see figure 5.3).

The market for personal computers can be divided into four segments in terms of use: home, education, scientific–technical and business. In this chapter we concentrate on the first three; the business/professional market, by far the biggest (see figure 5.4), is dealt with in more detail in chapter 7.

The leading makers of personal computers are battling it out for a slice of the booming business market, but it is through sales for home use that the personal computer has become most visible. Sales of home computers mushroomed in 1982–3, but intense competition and a slackening of demand led to price-cutting and a shake-out in the industry in 1984–5. The education market is growing, but remains small. The less visible scientific market, requiring more specialized products like high-

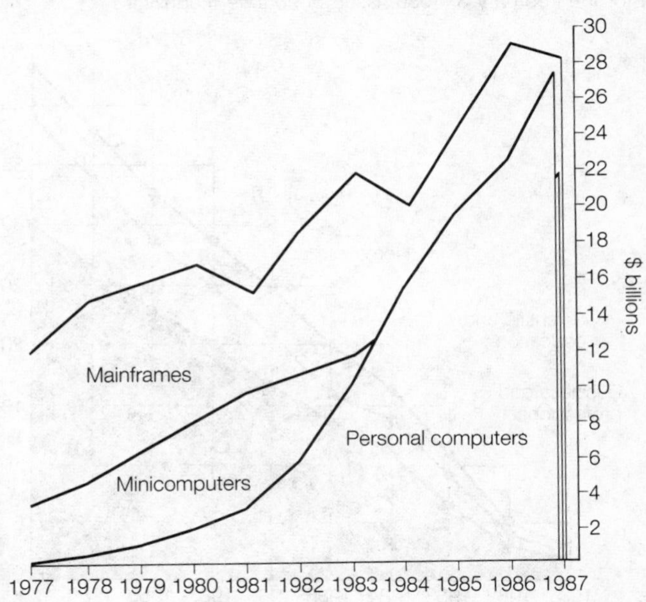

Figure 5.2 The meteoric rise of the personal computer: worldwide computer shipments by type, US manufacturers.
(Source: International Data Corporation. Data is from a paid advertising section prepared for the January 23, 1984, issue of *Fortune* magazine.)

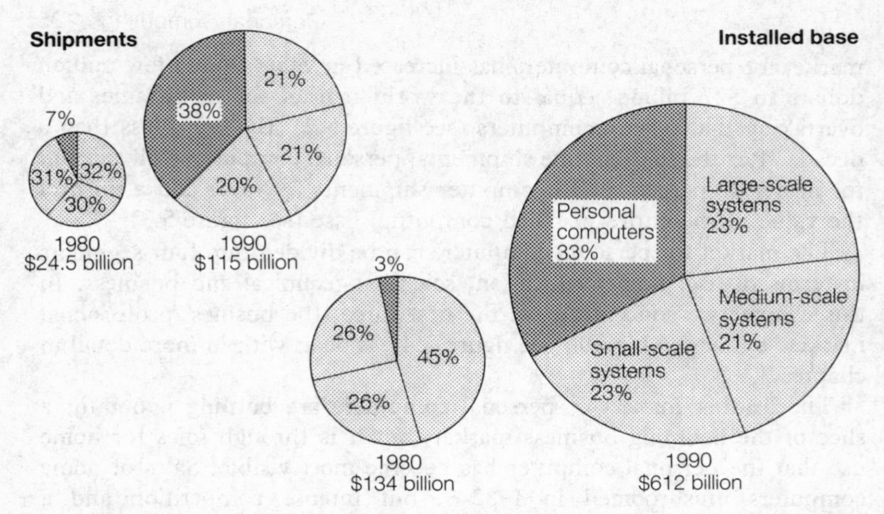

Figure 5.3 The personal computer starts to dominate: worldwide system figures, US suppliers.

(Source: International Data Corporation. Data is from a paid advertising section prepared for the February 3, 1986, issue of *Fortune* magazine.)

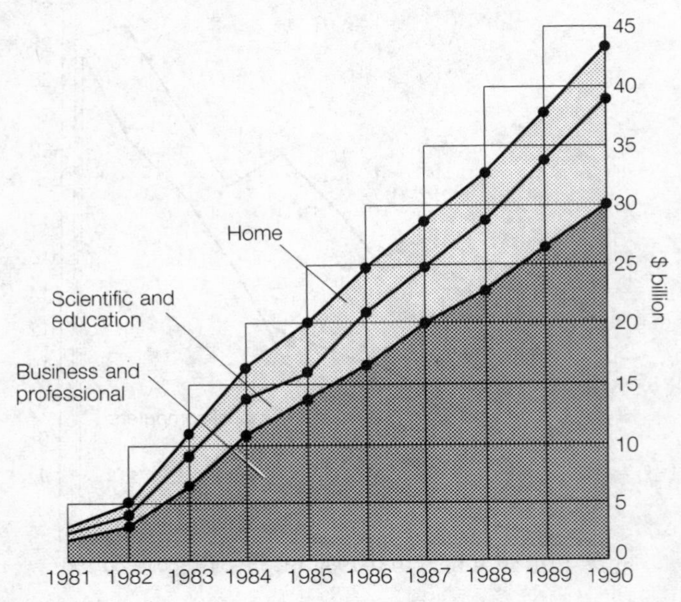

Figure 5.4 The uses of personal computers: worldwide shipments of personal computers, by application.

(Source: International Data Corporation. Data is from a paid advertising section prepared for the February 3, 1986, issue of *Fortune* magazine.)

performance "technical workstations," is actually much larger and is growing fast. Aided by price cuts, US sales in this market zoomed by 50 percent in 1985. A fierce marketing battle developed between longer-established workstation manufacturers, Apollo Computer and Sun Microsystems, and the heavyweights, in the shape of Hewlett–Packard, Digital Equipment and especially IBM. Analysts were predicting an eclipse of the two heavenly bodies.

Personal computers are now mass-marketed in the same way as soap powder or soft drinks and are sold off the shelves like cans of beans. Millions of dollars are being spent bombarding the public with TV, newspaper and trade press ads which seek to build loyalty to particular brands. Since the masses are unlikely to understand the technical details of an individual machine, marketing and distribution have become critical to a manufacturer's success. IBM spent millions on the highly successful launch of its PC model in 1981 with ads featuring a Charlie Chaplin look-alike. Apple had to match it, and soon after recruited from Pepsico marketing expert John Sculley, creator of the "Pepsi Generation" campaign. Hewlett–Packard has used butterflies in its ads, Commodore featured an "Einstein," and others have employed actors and pop stars to help shift the plastic.

Critics, however, allege that home computers in particular have been oversold: new owners have found it hard to make them do anything apart from play games. How many of the cheaper models are still in use months after purchase is anybody's guess.

The coming of the personal computer has also created a completely new form of retail outlet – the computer shop. Boyd Wilson and Paul Terrell opened the first full-time computer store, The Byte Shop, in Mountain View, California, in December 1975, and since then the number of computer shops has grown to about 15,000 worldwide. Some US cities have as many as 300 specialized computer stores, apart from all the other outlets selling computers. Franchised retail chains like Computerland have been a great success in the USA, as have manufacturer-owned stores like Tandy's Radio Shack and the Xerox stores.

But the 1985 turmoil in the personal computer business led to a shake-out among computer stores. A wave of price-cutting and chapter 11 bankruptcies swept through computer retailing. From a peak of 4600 stores in early 1985, the US total fell to 4100 in early 1986, of which the now-troubled Computerland had around 800. While newcomer Businessland grew rapidly to 140 stores and appeared to be thriving, IBM announced a moratorium on new dealer applications and Apple sliced 600 off its total of 2600 dealers.

A major new trend also became apparent as the well-heeled "Baby Bells" started moving into computer retailing. Bell Atlantic snapped up the 53-store chain, CompuShop Inc., for a modest $21 million. Pacific Telesis also announced plans for a chain of computer/phone shops. Nynex started its own chain, Datago, and in April 1986 purchased 81 IBM Product Centers. As a result, analysts were predicting that the industry would gradually consolidate into five or six major national chains.

In Europe, however, computer stores have not enjoyed anything like the success they have enjoyed in the USA. Prospective purchasers have continued to go to existing office equipment and consumer electronics suppliers or other sources. There has also been some skepticism about, and prejudice against, the concept of franchising. By the end of 1983, for example, Computerland had 571 franchised stores in the USA but only 38 in the whole of Europe. IBM had just 31 in Europe in 1986. In Britain, a number of home-grown chains like First Computer and Interface Network have been started, but they too have been slow to take off.

Publishers of magazines and books jumped onto the personal computer bandwagon soon after it got rolling, and in doing so started a publishing sensation. In 1981 there were less than a dozen computer magazines plus a clutch of old-established electronics journals. The following two years brought an avalanche of magazines devoted to personal computers. Despite the traumas of 1985, over 800 were available in the world by 1986. Newsagents devoted acres of display space to them at the height of the boom. The top 25 sold 66 million copies, and the top five boast circulations of around the 500,000 mark. Many are aimed at users of a particular machine – so their fortunes tend to rise and fall rapidly in line with hardware sales. As one analyst put it, "This market's so fast, it's a blur."

The rise of the personal computer has been one of the big surprises in the information technology revolution. It has placed considerable computing power in the hands of individuals – in particular, kids – and has enabled millions to tinker with computers and analyse such things as financial problems in their own homes. Eventually, the personal computer will enable many more people to work at home. People have also taken their personal computers into work, and this has upset the grandiose plans of certain corporations to sell fully integrated "automated office" systems. Increasingly, personal computers are being seen as the basic building block of office networks, manufacturing systems and educational/scientific facilities. With so much cheap hardware around, almost anything could happen.

The Personal Computer Boom

Looking at the personal computer market has been likened to observing the game of rugby for the first time. There's lots of pushing and shoving in evidence, but it's difficult to make out exactly what's going on.

The most recent chapter in the personal computer story opened in August 1981, when IBM's long-awaited PC emerged from the company's product development lab at Boca Raton, Florida. Apple, in a famous ad, cheekily "welcomed" IBM's entry into the personal computer market, which hitherto had been dominated by Apple, Commodore and Radio Shack's Tandy. It was clear right from the launch that IBM would be a major force in personal computers. Although based on a 16-bit microprocessor (the Intel 8088), the IBM PC offered nothing very new in the way of technology and was relatively expensive (costing upwards of $1500, and more like $4500 for a full system). But it had IBM's marketing clout behind it, and the PC quickly grabbed a large slice of the market.

Early in 1982, Tandy responded with a new version of its TRS-80 model, which incorporated a fully fledged 16-bit system and was faster than both the Apple II and the Commodore Pet. As corporate giants like AT&T, ITT, Xerox, Ericsson, Philips and Olivetti announced their own projects, the rush to get into personal computers began to resemble the

Klondike. Personal computers were sold for the first time in supermarkets, the business market (immediately dominated by IBM) was booming, and toward the end of 1982 a price war broke out as Texas, Timex/Sinclair, Commodore and Atari slashed prices. One Texas model, used as a "loss leader," dropped from $1000 to $199 in a few months. In the runup to Christmas, displays of personal computers quickly took over toy departments and whole new sections of stores were devoted to them. The personal computer business had become a bonanza.

In January 1982, IBM launched its PC in Europe. It was made in Scotland and marketed from the UK, which was in the midst of its own personal computer boom, boosted by new products from Britain's Sinclair and Acorn companies, based in Cambridge. Aided by homegrown entrepreneurial talent, the English language and the British government's scheme to put a microcomputer in every school, the UK rapidly became Europe's biggest personal computer market. Manufacturers like ACT, Dragon, Torch, Oric, Camputer and Grundy proliferated, although some were soon to bite the dust. By the end of 1982 it was said that 6 percent of UK homes had a personal computer – twice as many per head as in the USA and the highest penetration rate in the world.

In January 1983, Apple at last responded to IBM's challenge with the US launch of Lisa, a pricey ($9995) system based on a 32-bit chip which featured a remote control device called a "mouse" and elementary pictures on the screen called "icons" for ease of use. It also included "integrated" software and six built-in business programs. With Lisa, Apple hoped to end its six-year reliance on Apple II after the Apple III had proved such a disastrous flop in 1981. But, despite the fancy features and critical acclaim from the cognoscenti, high-priced Lisa failed to sell in large numbers, and attention shifted to the lower end of the price range, where makers of home computers and video games were engaged in a fight to the finish. By the summer of 1983 the price-cutting war had become intense, and it was clear that Mattel's Intellivision and Coleco's ColecoVision were in trouble. Atari and Texas Instruments reported stunning losses as Tandy and Commodore tightened their grip on the home computer market.

Meanwhile, IBM was rapidly consolidating its position in the market, especially the business sector. In October 1983, a *Business Week* cover story proclaimed IBM "the winner," with 26 percent of the entire personal computer market, followed by Apple, Tandy, Hewlett–Packard and DEC, in that order. In just over two years, and despite supply shortages, IBM had sold 750,000 PCs. The secrets of its success included the 16-bit design, low-cost manufacturing, multichannel distribution, aggressive pricing and, perhaps most important of all, its wide-open software, which encouraged software writers to produce programs for it. IBM had also taken on board the lessons of Silicon Valley and set up a wholly separate division to develop the PC, which operated like a start-up company. Introduced just as sales of personal computers were shifting from hobbyists to professional users, potential buyers were faced with a bewildering array of makers and turned with relief to IBM. And, as the saying goes, "Nobody was ever fired for buying IBM."

The success of IBM's PC created an entirely new industry in PC-compatible products, from programs and memory devices to look-alike computers. These "PC clones," from start-up companies like Eagle, Columbia, Compaq and Corona, appeared because IBM had published design details of the PC in order to encourage software writers – and because the company was unable to keep up with demand. The most successful was the Texas Instruments spin-off, Compaq, which made a PC-compatible portable that sold like hot cakes. In 1986 Compaq officially became the fastest-growing US corporation ever, when it entered the *Fortune* 500 more quickly and at a higher position than even Apple had achieved.

Despite IBM's pre-eminence, by the end of 1983 there were at least 150 US manufacturers of personal computers – double the number of 1981. Producing personal computers was originally an easy-come easy-go world in which firms like Vector Graphics, Victor Technologies and Fortune Systems came and went. Osborne Computer, founded by computer gadfly Adam Osborne, launched the first "transportable" computer in 1981 and notched up sales of $100 million in 1982. But in September 1983 Osborne filed for bankruptcy. With too many companies chasing a decreasing share of the market, a major shakeout loomed. And there were other signs that the industry was reaching maturity: the costs of entry were rising, and marketing and distribution were becoming more important than the technology.

In an attempt to further consolidate its position, IBM in November 1983 unveiled its cheaper home computer code-named "Peanut" – the PC jr, priced at $669. Again, industry specialists pointed out that the machine had nothing really new to offer and the Commodore 64, for example, offered better value for money. This time the critics had an effect on sales, which remained sluggish. Even so, Apple responded in January 1984 with the Macintosh, a kind of cheaper version of the Lisa. In the same month, British inventor Sir Clive Sinclair announced his QL (for "Quantum Leap") machine for the business market. Like Lisa, the QL used a 32-bit chip and at around $3000 was a lot cheaper. But it came too late to stop Timex from withdrawing from the US computer market in February 1984, after plummeting sales of the TS 1000 (or Sinclair ZX 81) model.

About the same time, the growth in sales of personal computers started to slow down, with the result that "only" about 5 million home computers were sold in the USA in 1984. Texas followed Timex in pulling out of the home computer market as losses mounted and as consumers tired of "playing games" on "toy" computers, or postponed their decision to buy. This left Commodore with about half the small computer market, a dominance reinforced by Coleco's shock decision in January 1985 to abandon the Adam computer altogether. Many consumers were also moving "upmarket," to Tandy, Apple and IBM machines that could now be had for less than $1000; the share of the total personal computer market held by machines costing less than $500 fell from 51 percent of sales in 1983 to 23 percent in 1985.

Supply started to outrun demand in the business sector, too: reports of a thriving "gray market" in discounted IBM PCs and IBM's launch of the more advanced PC/AT spelt trouble for the PC clones. Sales of IBM's PC jr at last picked up in 1985 and were nibbling away at the market for the Apple IIc. At the same time Apple, making a determined assault on the office automation market with the Macintosh and the Macintosh XL (Lisa renamed), faced a further threat from Commodore and a revived Atari under ex-Commodore boss Jack Tramiel, who took the 1985 Las Vegas consumer electronics show by storm with a Macintosh look-alike, the ST.

In early 1985 AT&T tried again to enter the personal computer market with the launch of its Unix PC. Designed to challenge IBM's PC/AT, it did surprisingly well later in the year, lifting AT&T's market share. IBM stopped making the PC jr after its market share slipped to 4 percent. This added to the number of computer "orphans" – those products no longer made by Commodore, Texas Instruments, Coleco, Timex, Mattel, Franklin and so on – although it did give a boost to "orphan user groups," which club together to help each other out with spares and repairs.

Atari's ST was finally made available at $1100, and the long-awaited Commodore Amiga eventually hit the market in August 1985 at $2100. But both faced problems getting retailers to stock them, and both lacked software support. While sales of the ST languished, Commodore bet everything on the Amiga in a bid to move upmarket to attract upscale and business customers. But despite some success and the acknowledged technological superiority of the Amiga's chip-set graphics and stereo sound, there was continued speculation about the future of Commodore.

In 1986, prices for most IBM-compatible personal computers in the USA dropped below the $1000 mark, confirming that the cost of each unit of p.c. computing power had dropped by about 30 percent per annum in the first half of the 1980s.

In the UK, too, personal computers were getting cheaper. Market leaders Acorn, Sinclair and Apricot slashed prices in early 1985, but this did little to help the struggling UK companies. In February 1985 Acorn hit severe financial difficulties: Italy's Olivetti paid off the company's debts in return for a 49 percent stake. In July Olivetti baled out Acorn again, raising its stake to 80 percent of a company now dubbed "Acornetti." Despite the success of Apricot's Zen model, there was a big question mark over Apricot's future after US losses and European failures. In fact, Apricot, Acorn and Sinclair all entered – and rapidly left – the US market after losing considerable sums.

But the most spectacular turnabout involved the fortunes of Sir Clive Sinclair, whose company was valued at $136 million in April 1985. Sinclair faced mounting losses through 1985 as sales fell. A well-publicized rescue by Robert Maxwell failed to materialize, and Sir Clive's C5 electric car was a gigantic flop. Twelve months later, in April 1986, Sinclair's computer business was sold lock, stock and barrel to British entrepreneur Alan Sugar for just £5 million. Sugar and his Amstrad company had caused a sensation in September 1985, when he launched the PCW 8256 word processor computer, retailing at just £399 plus tax. Sugar sold 500,000 units in less than six months, grabbing 20 percent of the European personal computer market overnight (IBM took exactly 33.3 percent, Olivetti 10.5 percent and Apple 9 percent in 1985) and making himself a fortune.

The Japanese, meanwhile, were preparing their own assault on the personal computer market. They were widely expected to join the fray as early as 1980–1, but it was not until 1984 that the first PC clones by Matsushita, Sanyo, NEC and Toshiba arrived, and they weren't a great sales success. In 1985 there came the long-heralded invasion of MSX machines, home computers that use the standardized operating system developed by Microsoft called Microsoft Extended Basic (hence MSX). In Europe and America, there was a lot of speculation about whether MSX would take over – just as the Japanese VHS format won out in VCRs – or whether the Japanese had produced "too little, too late." Some critics pronounced the Japanese machines "dead on arrival."

Technically, the MSX machines are not very advanced, but the great advantage of standardization is that programs and peripherals can be switched about and the consumer isn't stuck with obsolete equipment.

Philips of the Netherlands has already signed up for MSX, and there is talk of MSX being the standard for consumer electronics products like hi-fi rack systems and "smart" TVs.

The personal computer boom has also brought us the latest executive toy, the portable computer. Portable computers were originally souped-up hand-held calculators, but now they divide into what are called "luggables" (briefcase-sized computers, also called "transportables") and "lap-tops" (notebook-sized machines, also called "lap-helds"). Portables can be used for a variety of purposes such as word processing, electronic mail, data collection, field engineering calculations and mobile database queries.

The first portable computer – a 24 lb luggable – was the Osborne, unveiled in late 1981, and this was closely followed in 1982 by Grid Systems' Compass and Epson's HX-20. Compaq and Kaypro came out with PC-compatible portables, but the first real winner was Tandy's Model 100, a lap-top launched in 1983. In the rush to get on the bandwagon were start-ups like Gavilan Computer and Convergent Technologies, whose Workslate was the most sophisticated lap-top available. Sales of portables in the USA rocketed from 90,000 in 1983 to 470,000 in 1984, although they have not mushroomed since in the way predicted.

The technology that made portable computers possible was the invention of CMOS chips and, more especially, advances in LCDs (liquid crystal displays). These black-on-gray screens have rapidly developed from one-line slivers to 25-line by 80-character displays. The Japanese – through companies like Toshiba, Kyocera, Sharp, NEC, Mitsui and Epson – are world leaders, and currently there is a big research effort in Japan and the USA – through firms like Hamlin CrystalVision and Panelvision – to improve and perfect LCDs in order to take advantage of the growing portable computer market. Sales of LCDs are expected to leap from just $40 million in 1983 to $3 *billion* by 1992.

The Japanese lead is a result of earlier successes with watches and calculators. But this is a fast-moving technology, and there are many hurdles ahead in the race to make LCDs bigger, clearer and faster. Color LCDs are expected soon – this will make possible pocket color TVs – and LCDs will be used increasingly for projection devices and display panels in, for example, airplanes and cars. Economies of scale and technical innovation will make this a big growth area and a major industry of the future. In time, though, two other technologies – thin-film electroluminescence and gas plasma (for large screens) – might become more important if LCDs can't overcome their disadvantages of contrast and viewing angle.

Once Osborne and the other start-ups had demonstrated a demand for portable computers, the "old" firms like Apple got in on the act. In April 1984 Apple launched a portable Apple IIc, weighing in at 7½ lb, while Hewlett-Packard came out with its 9 lb Portable. Epson and Sharp also introduced new models. Partly as a result, Gavilan went bankrupt in the summer of 1984 and Convergent withdrew its Workslate. In October 1984 Data General of Massachusetts came out with the Data General One, the first lap-top with a (flip-up) 25-line by 80-character display. But only about 15,000 were sold in 1985 – about a quarter of what was expected. In fact, sales of portables in general remained sluggish in the USA and Europe in 1985. Poor-quality screens and high prices deterred many customers; others questioned the need for portability anyway.

Finally, in April 1986, IBM launched its long-awaited portable, the PC Convertible – which at 12.2 lb was more of a luggable than a lap-top. Early reactions were mixed. IBM has aimed its portable primarily at the busy executive. But both IBM and the other major manufacturers clearly see a big market for portables in the future among engineers, architects, journalists, welfare officers, tax collectors, insurance brokers and a great range of traveling salespeople and salesclerks in stores. There is also a huge demand for mobile machines among the military.

Apart from bigger and better screens, additional features such as modems, built-in printers and memory disks are being offered to tempt customers. But despite some impressive sales successes – like Zenith's sale of 15,000 lap-tops to the US Internal Revenue Service – many analysts are now saying that portables will never become more than a niche market.

In the future, all personal computers are likely to become even more "user-friendly" and to offer speech recognition as a standard option. The keyboard will remain the dominant input device for many years to come. But in an effort to overcome the keyboard or terminal "phobia" which afflicts millions of people, personal computer manufacturers will make greater use of the hand-held mouse, touch screens and touch pads, which allow users to draw pictures or graphs on the screen. Voice recognition, bedevilled by false promises and past failures, is still some way off – although some recent successes using "second-generation" machines indicate that useful systems may be with us within a decade.

The very success of personal (micro)computers and the proliferation of desktop terminals is likely to lead to the abolition of the distinction between personal computers and other types of computer. For example, a new generation of "super-microcomputers" – more powerful than microcomputers, but at $5000 upwards much cheaper than minicomputers – is threatening to oust minicomputers from the higher-level commercial and technical markets. "Desktop computing" and "professional personal computing" have taken over from "office automation" as the new buzzwords in the office-of-the-future stakes. We are witnessing a convergence of mainframes, minis and micros – and perhaps soon it will make sense to talk only of "computers."

The Software Revolution

"Software" refers to the programs or sets of instructions that tell a computer what to do. Without software, a computer is just a lump of silicon and plastic. It is like a stereo without records, or a musical instrument without a score. As the cost of (increasingly sophisticated) hardware declines, software has assumed an even greater significance. Hardware is now simply a commodity; software is where the action and the money are. In fact, its rate of development is largely setting the pace of the information technology revolution.

In the early days, mainframe computer manufacturers provided the basic systems software and customers wrote their own applications software under the manufacturer's supervision. Much of it is still in place, running airline reservation systems, doing phone company accounts and so on. But on June 29, 1969, IBM decided to "unbundle" and sell hardware and software separately. That opened the floodgates and allowed firms like MSA and Cullinet to set themselves up as specialist software houses. Most software sold or leased today (by value) still goes into big mainframe computers used by the government, banks, insurance companies, the utilities and other major data processors.

But the real revolution in software came about with the rapid spread of personal computers in the early 1980s. This has put software into the hands of individuals at reasonable cost for the first time. Personal computers and the software for them go together: as personal computers proliferate, the demand for software soars, and as more software packages become available, the demand for personal computers soars. Indeed, without an adequate "software base" of compatible programs, a personal computer will fail in the marketplace.

The software industry has mushroomed from a cottage industry of amateur programmers in the late 1970s to a multi-million dollar business today. Sales of software packages for personal computers – a mere $250 million in 1980 – are expected to leap from nearly $2 billion in 1984 to around $8 billion in 1989. The overall software market will grow from about $10 billion in 1984 to maybe $35 billion in 1989. In the space of a few years, the software industry has developed sophisticated distribution channels and has spawned such new figures as software publishers, agents and even software pop "charts" with software "stars" – like young Bill Budge, who made a fortune from such video game hits as "Raster Blaster." Today, it is estimated that there are around 2000 software companies and about 30,000 software packages available in the USA.

Systems Software

There are basically two kinds of software: systems software and applications programs. Systems software, or "operating systems," are the "housekeeping" programs that enable the parts of the computer such as the CPU, the memory devices and the printer to work together. The first popular operating system was CP/M, developed by Digital Research, but this was supplanted by Microsoft's MS-DOS after it was adopted by IBM for use in its PC, launched in 1981. By 1984, MS-DOS was beating CP/M six to one in the market. But MS-DOS is itself threatened by Unix, a more flexible system first developed by AT&T's Bell Labs some years ago and only now being adopted by a growing number of computer makers, such as Hewlett–Packard.

Unix was written for use by experienced programmers of minicomputers, and it attained a strong cult following in universities and research labs. Digital hit back in 1985 with a new system, Concurrent DOS, which is compatible with the software written for IBM's souped-up PC/AT. However, the main battle seems to be between MS-DOS and Unix. Some have billed this as a potential clash of the titans – pitting IBM with MS-DOS against AT&T with Unix. But already it seems that IBM has accepted that the "portability" (facility enabling it to be used on different machines) of Unix is a major advantage and has developed its own version, called Xenix. This places a big question mark over whether AT&T's Unix System V will, in fact, become the standard Unix system.

During 1985–6, AT&T made continuing efforts to get Unix – its only real trump card in the computer game – adopted as the industry standard. Through a $3.5 million press advertising campaign and a series of new product announcements, AT&T emphasized that Unix's portability meant that users would not be locked into one supplier. But industry analysts pointed out that Unix would not sell solely on its strength as an operating system – a range of good Unix-based applications software was needed as well.

Programming languages make it easier for people to communicate with machines by translating instructions into strings of 0s and 1s. There are a dozen or more languages, each designed for different users and applications in mind. The first popular one, FORTRAN (for FORmula

TRANslation), was developed by IBM in 1956 and used mainly on mathematical and scientific problems. BASIC (Beginner's All-purpose Symbolic Instruction Code) was written by Dartmouth professors John G. Kemeny and Thomas E. Kurtz and has been widely adopted in schools, colleges and personal computer programs because it is easy to learn and use. PASCAL, named after the seventeenth-century French mathematician, is difficult to use but precise. COBOL (COmmon Business-Oriented Language) is still the most widely used programming language for mainframe computers.

Two languages currently making waves are ADA, the standard adopted by the US Department of Defense, and LISP (for "List Processor"), which is fast and allows programmers to link symbols in ways that AI researchers believe is similar to the way the human brain works. ADA is named after Augusta Ada, the Countess of Lovelace and daughter of Lord Byron, who worked with Cambridge University's Charles Babbage on his 1834 "analytical engine." She punched the cards and is therefore credited with being the world's first computer programmer.

ADA is being heavily promoted by the Defense Department as its standard in an effort to save money. As a result, software start-ups are climbing on the bandwagon and producing new programs written in ADA. LISP, long the favourite language of the AI community, is now being acclaimed as the easiest and fastest language around – the computer language of the future, and the one most likely to break the so-called "software bottleneck." Some have even suggested that LISP machines will supplant Unix machines by the end of the decade as the sharp distinction between operating systems and programming languages becomes blurred.

Applications Programs

Applications programs turn the computer into something useful because they enable the system to tackle a specific task. These software packages allow the user to do such things as word processing, ledger accounting or mass mailing or can be tailored to the needs of a specific industry, such as banking, insurance, construction, distribution or medicine. VisiCalc, the business planning program, is the all-time bestseller among applications programs, with 700,000 copies sold since its introduction in 1979. Most of the programs sold to consumers during the personal computer boom were for playing games, but word processing, family financial planning and educational programs have since become more important.

The major trend in software today is the move toward "integrated software" – packages that permit the user to perform several different tasks because a number of programs are integrated together on a single storage disk. For example, the pioneer integrated software program, Lotus Development's 1–2–3, enables the user to carry out spreadsheet analysis, retrieve data from a database and display graphic material without having to change disks. The general idea is to more or less replicate a busy desktop on the computer screen – and this effect has been enhanced by the introduction of programs by such firms as VisiCorp and Microsoft which split the screen into "windows" displaying different types of material.

Makers of integrated software packages have become involved in a bitter battle for market share. Lotus heavily promoted first Symphony and then Jazz, which add word processing and communications functions to the calculating, graph-drawing and data-organizing functions already available on 1–2–3. Ashton-Tate, the new No. 1 in software for personal computers, responded with Framework and dBase III PLUS, and Microsoft brought out Excel. Meanwhile, Sidekick, an inexpensive desktop jotter designed by the young French mathematician, Philippe Kahn, has sold unexpectedly well.

The increasing sums being devoted to marketing software packages are a sign that the industry is reaching maturity. The "hard sell"' of software really began in 1982, and by 1985 huge sums were being spent on TV commercials and other attempts to establish "brand" names and images. For example, Lotus spent $1.7 on the promotion of Symphony for every dollar spent on R&D and no less than $7.5 million on its 1985 campaign to introduce Jazz. The software advertising industry has become a whole new genre in itself. Software is being sold by mail order, through bookstores and even in software supermarkets. Guides to software like the magazine *LIST* and publications like *The Whole Earth Software Catalog* help pilot consumers through the maze.

But it has not been roses all the way. In 1984–6 sales growth was much slower than predicted, and there were signs of a shakeout, as some software publishers went bankrupt and big name newcomers like *Readers Digest* pulled out of the market as quickly as they had entered. VisiCorp was embroiled in a battle with Software Arts of Wellesley, Massachusetts, which created VisiCalc in the first place. In late September 1984 the so-called "Software Quake" occurred: IBM announced it had developed 31 home-grown programs for business use and would enter the packaged software market itself. This was partly in recognition of the growing importance of network software – software that connects personal computers in a business setting. Other major hardware manufacturers plan to follow suit, and this will further intensify competition in an already very competitive market. IBM has the power to dominate if it wants to: some say it's just a question of how many software houses IBM will allow to survive.

Meanwhile, the Japanese have been pouring money into software R&D in order to produce "world class" programs. They face difficulties of language and cultural style but are determined to succeed. So far, Japanese firms have nibbled away at the video games market and specialist areas like banking and airline reservations, CAD and computer graphics. But generally the Japanese are weaker in software, and this is holding their computer industry back: Japanese machines lack packaged software and lose sales as a result. Standardizing on MSX is one strategy to get around this problem. How things shape up in the next few years depends very much on the complex politicking between IBM, MITI and the US and Japanese governments.

Because software has been so expensive, copying programs has become a major growth industry – creating the new and widespread phenomenon of "software piracy." In schools, colleges and computer clubs, youngsters run off duplicate programs for their friends or for resale – just as you would make copies of a cassette tape. Software rental agencies have mushroomed, with no questions asked about what customers do with the programs once they get them home. Some software companies put special codes or scrambler devices into programs to prevent illegal copying, but the codes are easily broken and scrambling techniques can be a nuisance to legitimate users.

In the USA, where software piracy costs the industry untold millions of dollars a year, major software houses organized a fightback to curb the problem. Micropro International, creators of the best-selling Wordstar word processing program, slapped suit on American Brands, the US tobacco group, for unauthorized copying and on United Computer Corporation, a fast-growing software "rental" agency, while in September 1983 the US Court of Appeals in Philadelphia extended the copyright laws to computer software in the case of Apple Computer vs Franklin Computer. Lotus sued Rixon Inc. and four other companies for copying 1–2–3. Some software manufacturers got together to form the Software Protection Fund, taking the view that informal pressure on major users was more likely to be effective in the long run than heavy-handed legal action. In 1986 IBM began to tackle the overseas piracy problem through court action in Singapore and Taiwan, the counterfeiting capitals of the world, where copying software is big business.

In Britain, various devices or "dongles" with names like Padlock and Copylock have been marketed in an effort to combat piracy, estimated to be worth £150 million a year. In 1984 the Federation Against Software

Theft (FAST) was set up by leading computer companies. In 1985 a parliamentary bill promoted by William Powell MP was finally passed, which extended the terms of the UK Copyright Act 1956 to computer programs, although skeptics were unsure whether it would prove workable.

Software in future is likely to be easier to use and more intelligent, and is likely to come as part of the hardware of a machine. Companies like Apple, AT&T and IBM – through their "usability" labs – are pouring resources into developing more user-friendly software, heralded by the appearance of menus, pictograms and windows on the screen. Network software linking personal computers, software linking micros to mainframes and sophisticated file management software or so-called database management systems are set for rapid growth. So are so-called expert systems software like Lightyear and Expert Choice, which aid management in decision-making.

Apple, which pioneered the use of the "mouse" and icons with the Macintosh machine, has helped spur a revolution in computer graphics with MacPaint, which works on the Macintosh. New software means that Apple Macs can be used not only for graphics but also for typesetting and full page make-up. Linked to laser printers, Apple Macs make possible in-house or "desktop" electronic publishing systems.

Taking Computers Home

The creation of the personal computer means that individuals can for the first time keep powerful computers in their own homes. Like TVs and telephones before them, home computers have rapidly become part of the domestic furniture. Soon, it is claimed, every home will have one.

The term "home" computer was first applied to systems – keyboard, CPU and printer – costing less than $1000, but nowadays this description is reserved for computers costing less than $500. The world market, which was worth $1.5 billion in 1983, had grown to $4 billion by 1986. But prices were tumbling, even though the machines were getting more powerful all the time. The installed base of home computers reached about 6 million in the USA and 2 million in the UK by 1985. The market leaders were Commodore, Atari and Tandy – after Timex, Texas Instruments and Coleco withdrew – but the position was changing very fast.

Who buys them? The typical owner of a home computer is male and in his early teens. Ownership is highest in households with such children. Most have been bought by youths in order to play games, but many have been purchased by concerned parents, anxious that their children do not "fall behind." There is also a significant market among adults who feel they need to learn about computing: in Britain – which proportionately has more computers in homes than any other country – a 1984 survey found that two-thirds of purchasers had bought home computers simply to "learn" about computers and computer programming.

In the USA, Gallup, Roper and Yankee Group surveys in 1983–4 confirmed that between half and three-quarters of households used their computers primarily for playing games, but a sizable proportion said they were using them for word processing or to do office work at home, as a learning tool for children and parents, for information retrieval by tapping into databases like CompuServe and The Source, and for balancing checkbooks and the household budget. A few were using them to store recipes and addresses and to count calories. In the 18–29 age group, Roper found that 25 percent of non-owners expressed interest in owning a home computer, but this declined to 16 percent among 30–44-year-olds and tailed off to a tiny 3 percent among the over-60s.

While game-playing on computers was enjoying a boom, there was an underlying concern among manufacturers and analysts that home computers had been "oversold" on the basis that "you didn't know you needed one until you got one." Many customers, faced with the problems of operating the new machines and manuals written in gibberish, became rapidly disillusioned. Potential buyers did not perceive a real need for a home computer and were unsure about what they might do. Strenuous efforts were being made by the industry to make home computers more useful and easier to use for the great mass of people, not just hobbyists. One suggestion was that home computers could become the core of sophisticated home control systems, which would handle security and monitor heating, lighting and ventilation.

Predictions are difficult in this area, but two trends seem apparent. First, we may be witnessing the merger of home computers and the telephone – thanks to new, low-cost modems, which convert a computer's digital information into analog signals for transmission over the telephone lines. Thus, home computers will be used more and more for on-line information (especially financial information) retrieval, informal contact networks and for home banking and shopping.

Second and more important, we are witnessing the convergence of "home" and "business" computers, as the price of "business" models from Apple and IBM comes down below the $1000 barrier and better-off home users move "upmarket" to meet them. Sales of the more expensive home computers have increased relative to models at the lower end, which are increasingly perceived as "toys" just for "playing games." Buyers of the most sophisticated machines tend to be older and better educated; they have higher incomes and are interested in office-type uses – especially word processing – in the home, perhaps as a prelude to homeworking.

Computers have not been welcomed into every home, however. Easy access to hardware has created the new social category of "hackers" – compulsive and obsessive programmers (sometimes called computer "junkies" or "computerholics") who get hooked on computers and spend all their waking hours in front of the flickering screen (see also chapter 9). There are fears that the new machines might inhibit or distort the social development of the young. Wives whose husbands seem to have fallen in love with their computers have become self-ascribed "computer widows," and there have been reports of families actually breaking up in the face of the computer invasion. As one divorcee put it, "Who would have ever thought that I'd lose my spouse to a machine?"

Psychologists set to work on the problem early on, when home computers first arrived. Dr Neil Frude of University College, Cardiff, in his book, *The Intimate Machine: Close Encounters with the New Computers* (Century, London, 1983), showed that hackers develop through three distinct stages. First, their enthusiasm for computers is matched by a lack of interest in everything else. Second, the hacker's normal pattern of life becomes disturbed – sleep and mealtimes become irregular. By the third stage the hacker is characterized by "profound social withdrawal" – an inability to communicate with others apart from other hackers, or to show emotion, except when it is to do with the machine. The difference between the computer hobbyist and the hacker, Frude argued, was akin to that between the social drinker and the alcoholic.

Sherry Turkle of the Massachusetts Institute of Technology sees personal computers as a projective medium upon which people stamp their own personality, rather like the famous Rorschach inkblot test. But even more importantly, she argues in *The Second Self: Computers and the Human Spirit* (Simon and Schuster, New York, 1984), computers are heavily influencing the way people think about themselves, others and society at large. Hackers, she says, lock themselves away from the world and inhabit their own little safe and controllable "microworlds." Drawing on Piaget and Papert, she also claims that the new generation of "microkids" are even beginning to think of themselves as machines rather than people.

But Turkle rather overstates her case, ignoring the fact that other obsessive hobbyists – like radio hams – have been around for years. Unlike Frude, she seems also to forget the positive side of computers.

Many self-taught hackers have become expert programmers and brilliant software engineers – in some cases, after having been branded "failures" and written-off by the school system: computers have given them a new confidence. Computers can bring people out as much as fence them in.

Hackers put in long hours, but whoever heard of a successful ice-skater or brilliant musician who didn't become obsessive and put in hours of daily practice? Hackers have also been credited with some remarkable feats: for example, Peter Leppik, aged 15, of Golden Valley, Minnesota, hit the headlines in 1985 when he cracked the code on a program owned by a suspected child molester. Not so widely praised was a British hacker who broke into the Duke of Edinburgh's secret electronic mailbox on Prestel and left a rude message.

Consumer Electronics: the Video Age

Computers in the home are just part of the wider consumer electronics bonanza of the postwar era. First, of course, there was the TV set: in 1951 only 8 percent of US homes had one, but penetration reached over 80 percent a decade or so later. Then came cheap record players and transistor radios in the late 1950s, stereo hi-fi and cassette decks in the 1960s and video game systems in the 1970s. Home computers followed, and more recently there has been explosive growth in the number of video cassette recorders (VCRs) in homes. Videodiscs and miniature TVs have not yet become such hot sellers, but compact disc players and remote control, stereo, digital, high-definition, flat-screen and big-screen TVs are catching on fast. Judging by the overall expenditure figures, the world's love affair with the automobile has a close rival in consumer electronics.

In fact, there is talk of fully integrated domestic entertainment and control systems – even home robots – by the early 1990s. Someone coined the ugly term "infotainment" to describe the process by which domestic audio, video, telephone and computer technologies are merging to create a new generation of sophisticated home electronic products. The arrival of the digital or "smart" TV – which chops the incoming analog signals into digital data – is seen as the key to all this, because these new, computerized TVs will become the central communications hub or "command center" of the modern high-tech home. That is why the Japanese are standardizing on MSX as a part of their strategy for dominating the consumer electronics market.

Video Games

Video games pioneered the home computer invasion and for a while were one of the most profitable industries ever. In the beginning there was "Pong," the simple black-and-white "tennis" game introduced in 1972. Its inventor, Nolan Bushnell, founded Atari in the same year but sold out to Warner Communications for $28 million in 1976.

By the end of 1977 consumers were becoming bored with "Pong"-type games, but the invention of the programmable chip saved the day for Atari. Its Video Computer System was the first of a new generation of video games, which enabled the player to switch cartridges, just like a cassette tape recorder. Although introduced in 1977, sales did not take off until 1979–80, but when they did they went like a rocket. Atari's revenues shot up from $3 million in 1973 to $770 million in 1981, with profits leaping from $1 million to $160 million over the same period. When Bally came out with "Space Invaders," a coin-operated game designed by the Japanese firm, Taito, the market went crazy: video games became the most sensational craze of all time. Sales of video games cartridges with names like "Missile Command" and "Asteroids" rapidly ran into millions, video game arcades sprouted up everywhere, and youngsters competed to see how long they could play for – the 1981 winner was Steve Juraszek, aged 15 of Arlington Heights, Illinois, who managed to keep a 25 cent game of "Defender" going for 16 hours and 34 minutes. Doctors identified new complaints like "Space Invaders elbow," while social psychologists intoned that figures like Atari's "Pac-Man" – the all-time best-selling game, with 7 million cartridges sold – were turning the young into morons.

In 1981 Mattel came out with Intellivision, a second-generation machine offering even more intelligence and sophistication, while in August 1982 Coleco launched ColecoVision, heralded as the "third wave" in games. But with firms like Activision and Milton Bradley (MB) toys rushing into the market, increasing competition and eroding profit margins, severe doubts were being expressed as to whether the boom could possibly continue. In addition, there were unmistakable signs that kids were beginning to tire of the craze.

The crunch came one week in December 1982. Atari's fourth-quarter forecast prompted Warner and Mattel's stock to slump on Wall Street and a crisis of confidence hit the industry. Despite Coleco's success in early 1983 with ColecoVision and "Donkey Kong" (which made the kids go ape), overall revenues were plummeting as video games faced new competition from home computers. Video games had generated revenues of $2.5 billion in 1982, but that proved to be the peak, and the figure fell to $1 billion in 1984. In August 1983 market leader Atari, in effect, went bust. This was an extraordinary turnaround in less than 12 months for the brightest star in the video game galaxy.

In the 1983 pre-Christmas season, the name of the (video) game became Survival. A reorganized Atari attempted a comeback with new games, but the industry was caught in a downward spiral of falling demand, over-supply and price cuts. Home computers were the "hot" product in the Christmas market, and in January 1984 both Mattel and MB decided to quit video games altogether. Cinematronics enjoyed a modest, temporary success with "Dragon's Lair," a new type of video game using a laser disk to generate movie-like animated pictures, but both Atari and Activision continued to rack up enormous losses. It was "game over" for video games – they had been well and truly zapped. "The trend," said one video arcade manager, "is back to pinball machines."

Video games were the precursor not only of home computers, but of an even bigger boom in video generally. Thanks to the video cassette and the portable video camera cassette recorder (CCR, or "camcorder"), there has been an explosion in the use of video for home movie-making and in education, training, publicity, demonstrating and selling everything from houses to cosmetics. We are into the Video Age, and the social consquences could be far-reaching.

VCRs

The device of central importance in the video explosion is the video cassette recorder (VCR), which has sold by the million in the USA, Europe and Japan. Not since the advent of the television set itself has a product made such an impact: the VCR population of the world already exceeds 100 million. Over 60 percent of Japanese, 46 percent of UK and 33 percent of US homes had one by 1986.

Sony introduced the first VCR in 1975, but it was expensive and slow to catch on. Two years later, Matsushita announced a cheaper version, which worked on a different format called VHS and also featured longer-running tapes than Sony's Betamax machine. Matsushita then outwitted Sony in the marketplace by offering VHS licenses to other companies keen to make VCRs, thus spreading the VHS base. As a result, VHS now has about 80 percent of the market and Betamax is down to 20 percent.

The Royal Wedding of Prince Charles and Lady Diana in 1981 did much to stimulate demand for VCRs in Britain, which became the world's fastest-growing market. Defying the recession, 800,000 Britons confirmed their love for the "telly" and lashed out £600 million on VCRs and tapes in 1981 alone, giving VCRs a household penetration rate of 19 percent by the end of 1982, compared with only 6 percent in the USA. Video projectors rapidly appeared in British pubs and working men's clubs, and more than 25,000 video shops opened up in the space of 18 months. Sales growth in Britain then slowed, just as it was picking up in Europe and America. In 1983, 4 million VCRs were sold in the USA; in 1984 – spurred by lower prices – 7 million were sold and in 1985, 12 million. VCRs became a hotter item than home computers as the Japanese manufacturers struggled to keep up with demand.

At the same time, the VCR bonanza sparked a remarkable boom in the sale of home videos – copies of feature films like *Raiders of the Lost Ark* (850,000 cassettes sold by mid-1984) and how-to-do-it videos like *Jane Fonda's Workout*, the second biggest seller with 600,000 copies sold. Music videos, especially *Making Michael Jackson's Thriller* (350,000 copies sold), also enjoyed amazing success, with teenagers using the family VCR as a kind of video jukebox. More recently, *Rambo* has topped the video bestseller lists. Sales of VCRs and cassettes fuelled each other, but not all cassettes among the many thousands available have sold in such large numbers.

Yet "time shifting" – recording items off the television for later viewing – is still the main reason why most people buy a VCR. According to a US survey, 77 percent of VCR buyers ranked the record-and-playback facility as "very important" and only 27 percent gave the same importance to renting pre-recorded cassettes. The average user recorded 6.9 hours of programs off the air each week – with regular soaps top of the list. Many users *preferred* taping shows for later viewing because it enabled them to fast-forward through the commercials – much to the chagrin of Madison Avenue.

The very success of videos has created the major problem of video piracy. All over the world, "pirates" have been doing a roaring trade copying films onto video cassettes for resale to suspecting and unsuspecting customers. In some countries of Africa and South America, the market for videos is almost wholly "pirate." In the USA, the Motion Picture Association of America started a worldwide campaign against piracy after the Supreme Court ruled that the home taping of broadcast material was legal – but that duplicating tapes was not. Nobody knows the true size of the US pirate video market, and it is generally thought to be in decline. But it has undoubtedly helped undermine demand for Home Box Office and other cable TV movie channels.

In Britain – dubbed the "piracy capital of the world" – video pirates hit the headlines when films like *E.T.* became widely available on cassette long before its cinema release. The pirates were estimated to have captured 65 percent of the British video market and a thriving export trade. As a result, the Copyright Act was amended in 1983 to increase the fines for piracy, and the Federation Against Copyright Theft (FACT) was set up under Peter Duffy, the former commander of Scotland Yard's anti-terrorist squad. After hundreds of raids, scores of arrests and the seizure of thousands of illicit tapes – thanks partly to a new technique of "marking" legitimate products – FACT was able to claim in 1985 that piracy had been steadily reduced to just 20 percent of the UK market. Even so, poor-quality pirate versions of films like *Ghostbusters* were a continuing nuisance.

The camera cassette recorder (CCR) promises to revolutionize home movie-making and could become as popular as VCRs. CCRs combine a video camera and cassette recorder into a simple, compact unit. They are small, lightweight and portable. JVC took an early lead in the market, using its half-inch-tape VHS format, but Sony responded in 1985 with the launch of its tiny 2.2 lb Handycam, which uses an 8 mm format. In a rerun of the VCR battle, the two incompatible technologies are fighting it out. As prices fall, sales rise and newcomers like Kodak and South Korea's GoldStar enter the market.

Videodiscs and Compact Discs (CDs)

The mega-flop of the consumer electronics/video boom has been the videodisc – its only rival for the title being the quadraphonic sound systems of the 1970s which failed to sell. Amid great excitement in the late 1970s, RCA, JVC of Japan and Philips of the Netherlands announced that they were pouring millions of dollars into the development of the videodisc, which was seen as the mass consumer product of the future. Videodisc players are the audio-visual equivalent of the gramophone – and videodiscs the video equivalent of LP (long-playing) records. RCA in particular spent huge sums on its SelectaVision system, which was finally launched in the USA in 1981. Versions of the Philips LaserVision system followed in 1982.

One obvious problem was that rival manufacturers had developed three completely incompatible systems. SelectaVision involves a diamond stylus that follows a tiny spiral groove in the silver platter, which has to be stored in a protective plastic "caddie." The LaserVision system uses optical disc technology, in which a laser beam "reads" billions of tiny reflective pits in the disc, which are protected by a permanent plastic coating. JVC's VHD system is a kind of hybrid; it uses a stylus – but the stylus doesn't follow a groove, but is guided by servo-control instead.

This created confusion enough, but skeptics including Sony argued that the world was not ready for another revolution in home entertainment after the success of VCRs. Videodisc players were cheaper and produced better-quality pictures, but they were unable to record broadcast TV programs and they generally lacked the flexibility of VCRs. There was also a shortage of film titles on disc and the discs were plagued with production problems.

Rising concern in the industry suggested that disaster loomed. RCA's system was a flop in the crucial 1981 Christmas market, selling only 105,000 against a projected 200,000 by year's end. Philips and Pioneer (selling their LaserVision-based system, called LaserDisc) sold around 75,000 between them. In Britain, Philips introduced lay-offs at its Blackburn, Lancashire, disc-making plant in 1982 and JVC (linked with Thorn–EMI) once again postponed its European and US launches, planned for 1983. Apart from steady sales to minority groups like "movie nuts," videodisc player sales stayed in the doldrums. Finally, in April 1984 came the bombshell: RCA was ditching SelectaVision altogether after racking up cumulative losses of $280 million since 1981 plus the $300 million development costs. This was despite selling a total of 500,000 players and 10 million discs. Other manufacturers, however, vowed to carry on, albeit on a limited basis and for specialist "niche" markets.

As a mass consumer product, videodiscs must be judged a failure. But they do have a future as an educational tool and in other areas where high-quality pre-recorded material is required. There is growing interest in "*interactive*" videodiscs, which respond to commands from the viewer, and other systems which link videodiscs with personal computers. IBM, for instance, recently said it was extending its use of discs in training and promotion. Videodiscs are also ideal for storing large amounts of information or large numbers of pictures in video "encyclopedias." JVC is developing an interactive version of its VHD system for the consumer market, while Philips's LaserVision is becoming the dominant system in the growing commercial and educational markets.

Ironically, it is the videodisc's younger brother, the compact disc (CD) which has achieved much greater success. CDs are basically an abbreviated 4¾ in. in diameter version of the laser-based videodisc. They have their sound recorded in the form of a digital code engraved as a pattern of tiny indentations on the metal disc. As the disc spins, a laser beam "reads" the code, which contains millions of digits and runs for 3.3 miles on each side of the disc. The code is translated back into signals to the amplifier. Since nothing but the laser touches the disc, there's no hiss, crackle or pop you get with records – and no wow or flutter you get with tapes. What's more, CDs will never wear out – the first play is the same as the one-thousandth.

Introduced by Philips and Sony only in 1982, worldwide sales of CDs reached 40 million in 1985. In the USA, CD players have become the fastest-selling machine in the history of consumer electronics. Aided by falling prices, sales of CD players in the USA zoomed from just 35,000 in 1983 to 240,000 in 1984 and 700,000 in 1985. Disc sales shot up from 770,000 in 1983 to 15 million in 1985. Their value: 12 percent of all record and tape sales – in two years flat. Apart from Philips of Holland, Japanese companies dominate the market.

Japanese companies are also set to dominate the emerging market for digital audio tape (DAT), which complements the switch from analog to digital in discs. DAT offers a single standard for all tape-recording, although some machines will be equipped with stationary heads (S-DAT) and others with rotating heads (R-DAT).

Future Possibilities

As to the future in consumer electronics, many argue that the key innovation will be the digital or "smart" TV, which will transform the humble TV set or "boob tube" into the central command post of the modern home. Digital techniques can not only improve sound and picture quality by "cleaning up" the incoming signals; they can also allow information such as pictures to be stored, manipulated – for example, by zooming or freeze-framing – and converted into other types of digital data. In 1983 ITT unveiled its Digivision system in the first production model of a digital TV set. Other manufacturers have brought out stereo TVs, and TVs with teletext and add-ons like remote control.

Meanwhile, flat TVs are on the way: in early 1985, for instance, Matsushita demonstrated its 4 in. deep screen and Sony has shown a 1125-line "high-definition" TV (HDTV) set which is a vast improvement on current 525- or 625-line models. HDTV makes possible the display of TV pictures on large (for example, 5 ft by 3 ft), wall-mounted screens. These would seem to have greater sales potential than tiny TVs utilizing LCD technology, which have so far failed to excite consumers. But there is a need to resolve the current international disagreement over HDTV standards, which came to a head at the four-yearly session of the International Radio Consultative Committee (CCIR) in May 1986, when the European nations blocked acceptance of the Japanese–American 1125-line/60 Hz standard.

Computerized home control systems remain a distinct possibility. General Electric (GE) of the USA – which has already found out consumers prefer the new digital controls and diagnostic options on their appliances – has developed a simple home controller called Home-Minder, which uses the TV set and telephone to turn lights and appliances on or off by remote control. GE's second-generation system, Homenet, will be higher-powered and will have a two-way capability, using sensors to feed back information about how the dishwasher is doing, and so on. The general idea is to link everything up in a way analogous to local area networks (LANs) in offices and factories.

GE's leading competitors in the USA are skeptical of the value of or demand for integrated home control systems, but they may change their minds when they realize that the Japanese are also working toward them. Using what they call the "new media" of cable, videotex, teletext and DBS to deliver information to and from the home, the Japanese plan to tie their interactive home information and entertainment systems together using four key technologies: *processing*, which will be based on the MSX standard; *man–machine interface*, using advances in touch screens and mouse-controlled iconic and window displays; *mass storage*, based on hybrid and interactive videodiscs; and *display devices*, such as flat-screen, large-screen and high-definition TV.

This could be the shape of things to come in the 1990s, while the much-talked-about home robots and automatons – which still provide endless fascination for numerous small development companies all over the world – probably won't arrive in useful form until the year 2000.

Working From Home: the Electronic Cottage

With more personal computers being taken home, new software packages available and the convergence of "home" and "business" computers, the scene seems set for a big increase in home working. The lonely pioneers of Alvin Toffler's "Electronic Cottage" are, it is claimed, ushering in a new Age of Telecommuting.

"Telecommuters" work wholly or partly from home, while the term "teleworkers" has been applied to those who work anywhere outside the office – in cars, hotels and airplanes as well as at home. Jack Nilles, Director of the Information Technology Program at the University of Southern California's Center for Futures Research, predicts there could be 15 million telecommuters alone in the USA by 1990. It is said that the first Industrial Revolution largely destroyed the cottage industries, taking people out of their homes and into the new factories: now the information technology revolution is enabling people to return to their homes.

The last two or three years have seen a big increase in working at home, although the trend predates the new technology. In the USA, where homeworkers now comprise 10 percent of the entire workforce, some 200 companies have recently initiated telecommuting experiments and another 40 or so have formal telecommuting programs. A new National Association for Cottage Industry has been formed to promote homeworkers' interests. In Britain, Department of Employment surveys show a steady increase in homeworking over the last decade, especially among white-collar workers, to 1.7 million or 7 percent of the workforce in 1985. One surprise: 47 percent of them are men.

The trend to *working* at home is also part of a wider and marked British trend to "homecenteredness" recently reported by the Henley Centre for Forecasting. People are staying at home more partly because of successive recessions and partly to enjoy their videos, food processors and other electronic gadgetry – and the British pubs, restaurants and cinemas are allegedly suffering as a result.

There are many different types of homeworker and homeworking, ranging from the on-line computer programmer who is continuously in touch with the office to the executive who occasionally takes work home. Blue Cross/Blue Shield of South Carolina, for example, has 30 or so "cottage coders and keyers" who process medical claims at home outside of normal working hours. The company has found that they are 50 percent more productive than office workers and the error rate has plummeted. Aetna Life & Casualty of Connecticut has a number of systems programmers working at home, who have notched up similar productivity gains. Control Data, Mountain Bell and New York Telephone are equally enthusiastic about their telecommuting experiences.

In Britain one of the best-known examples involves Rank Xerox executives, who have been "sent home" by the company since 1981 in a bid to cut office overheads. These "networkers," as they are called, have in fact been set up in business in their own right by Rank Xerox, who have provided regular work on a contract basis and all the latest equipment – so the experiment has a technical aspect as well as a sociological one. None of the 50 or so executives had dropped out of the program by 1986, and some had become very successful – although it is important to note they had all been carefully screened for suitability before being allowed "out." Apart from "networking," Rank Xerox is also experimenting with "neighborhood offices" some way from metropolitan centers where fellow freelancers can meet. This is also being tried in Sweden and France.

The main advantages of telecommuting are savings in time, money and stress through avoidance of the daily commute; savings in real estate, furniture and energy costs at the downtown office complex; increases in productivity achieved by workers in a familiar, stress-free environment rather than a busy office where distractions, constant interruptions and what Studs Terkel termed "schmoozing" (wandering around chatting with fellow employees) absorb a high proportion of working time; and the creation of new employment opportunities for mothers and the physically handicapped.

On commuting costs, Jack Nilles has calculated, for example, that if the USA turned just one in seven of urban commuters into telecommuters there would be no need for the country to import oil. The smog-creating urban automobile accounts for one-ninth of total US energy consumption, and commuting costs account for a growing proportion of salaries. On office costs, a British study found that the overheads for employees in central London amounted to 2.7 times their wages. US surveys suggest that frequently travelling personnel cost four times their salary to keep on the road.

The productivity changes reported so far in telecommuting experiments are all on the plus side and are rarely less than 35 percent. Perhaps the most extensive US study – of 1200 employees by Electronic Services Unlimited (ESU) of New York – found average productivity gains of between 40 and 50 percent. Many homeworkers claim to achieve in 25 hours at home what would take 40 hours in an office.

Telecommuting also helps those who are unable to hold down conventional jobs. The Lift project, for example, in Northbrook, Illinois, enables the handicapped to work at home as programmers, and American Express has a similar scheme for handicapped word processor operators, Project Homebound, in New York. The British government also has a trial project for the disabled. Thousands of women with young children have been able, through telecommuting, to continue in paid employment. More and more US companies are agreeing to homeworking, often in order to hold onto highly valued female employees who want to have children. In Britain nearly 1000 women programmers are working at home for a company called F International, which was set up by a group of female executives with domestic ties who also preferred to live outside the big cities.

Telecommuting can thus play a decentralizing role in society, bringing new hope to depressed regions and even the Third World. Jack Nilles says: "Since the theoretical locational freedom made possible by telecommuting is limitless, a telecommuting society could scatter dwellings more or less uniformly between its boundaries." An increase in telecommuting could thus have major consequences for transportation systems, urban planning and even the landscape. An increase in *international* telecommuting, of course, could mean the export of jobs to and from other nations.

The main disadvantages of telecommuting are that managers fear loss of control over their employees and the employees themselves are in danger of becoming isolated, getting out of touch and losing out on promotion. There are also the inevitable technical glitches and fears that homeworkers – who are less likely to be organized into unions – will gradually become exploited second-class citizens. Computerized home workshops of the future could be no better than the cottage industry sweatshops of the past.

Most of the studies so far indicate that technical problems are easily overcome – it is dealing with the people involved that is more difficult. Managers in particular have tended to resist change, not wishing to relinquish their traditional intra-office power. But their "loss of control" fears have not been confirmed: homeworkers are not only more productive, they also produce better-quality work *without* supervision, according to analysts like ESU. Although using freelance homeworkers is therefore more profitable for companies, there has been no desperate rush to introduce homeworking largely because of management resistance.

Employees rarely feel isolated in the ways predicted, although this is partly because most telecommuters so far have been self-selected and highly motivated. In other words, homeworking tends to suit some people and not others. As the Rank Xerox program shows, the increase in homeworking is expanding the ranks of these independent, self-employed freelance workers, who sell their goods rather than their time, and who are paid for their performance rather than their attendance.

That must be good news for any national economy, but there is serious concern that such gains may be offset by human costs. For example, an official British study by Catherine Hakim, of the Department of Employment, found that the new freelancers fared less well than their office-based counterparts in terms of pay, while holiday pay, sickness benefit and pension rights were virtually non-existent.

Another British study, by Ursula Huws for the Equal Opportunities Commission, uncovered widespread evidence of financial exploitation among her sample of 78 new homeworkers, most of whom were computer professionals. Although the majority said that they were happier working at home without interruptions, and that they gained in terms of an improved quality of life, nearly all had pay levels £2 or more an hour below the average for their grade in more traditional locations. Moreover, most had no job security, no legal protection and no trade unions to stick up for them. They were thus at the mercy of their employers who could pick and choose at a time of high unemployment.

This grim picture of high-tech homeworking hardly accords with some of the more utopian visions of life in the Electronic Cottage. But whatever the case, as the information technology revolution proceeds, homeworking is bound to increase quite considerably. Its limits will be set not by the technology involved, but by the human and socioeconomic barriers that will need to be overcome.

Computers in the Classroom

Computers are now common in the classroom – and not before time. As long ago as 1962, Control Data Corporation announced the Plato system, a revolutionary method of linking classroom terminals to a large, central computer for teaching purposes. But the market was slow to develop, and the long-heralded revolution in educational technology never materialized. During the 1960s and 1970s a lot of money was wasted on expensive, over-sophisticated computer systems and unused "audio-visual" (AV) aids.

The personal computer changed all that. With the shrinking cost of computer power, sales of personal computers for educational use took off in 1980–1 and didn't stop. Governments and companies rushed to put personal computers in every school, and parents panicked if their child's school hadn't got one. Computer manufacturers queued up to sell or give away their products, and book publishers transformed themselves into publishers of educational software. In 1984, 350,000 computers were sold to US schools and the educational software market was worth $250 million. The so-called K–12 (kindergarten through grade 12) market grew by another 30 percent in both 1985 and 1986.

There was much talk of "computer literacy" and the "keyboard generation." It was said that computer literacy would take its place alongside the 3 "R"s of reading, writing and 'rithmetic as the fourth key skill, and that computers in schools might soon replace pencils. *Time* magazine in 1982 ran a cover story on the remarkable phenomenon of "The Microkids" – the new computer-literate generation who were more at home in front of a VDT screen than a children's book and who knew more about computers than their parents – and even their teachers. Someone also discovered the "computer toddlers" and the "pre-school software" market, while "computer camps" (replacing summer camps) enjoyed a mini-boom – not just among the microkids but among adults, too: the French-owned Club Mediterranée, for example, opened 23 computer centers, combining sun, sand, sea and software.

The battle to sell hardware to US schools really began in earnest in 1982, when Apple's Steve Jobs persuaded California Congressman Pete Stark and Missouri Senator John Danforth to introduce a Bill allowing tax write-offs to manufacturers donating computers to schools. Jobs figured that his "Apple for the teacher" strategy would establish long-term brand loyalty and would also generate home sales of Apple products. The idea was scotched in the Senate finance committee by Senators Robert Dole and Howard Metzenbaum but was allowed in California, where a state tax credit was agreed worth 25 percent of the value of the hardware donated: by the summer of 1983, Apple had "seeded" California schools with 10,000 computers. This was followed by a 1983–4 campaign offering $2495 Apple Macs to schools and colleges for $1000. Tandy, Commodore and others all joined in – as did IBM after the launch of their PC jr. But in April 1986 Apple still had 54 percent of the K–12 market with Tandy on 18 percent, Commodore 16 percent and IBM just 3 percent.

In Britain concern that the nation was falling behind in the technology race became widespread after the 1978 screening by the BBC of the documentary, "The Chips are Down." By April 1981 the UK government was ready to launch a £4 million "Micros in Schools" scheme, part of a £10 milion-plus Microelectronics Education Program designed to put chips on the menu in schools and colleges throughout the country. The first Micros in Schools program offered half-price Acorn and Research Machines computers to secondary schools – and only 500 out of the 7000 or so schools failed to take up the offer. A second program offered half-price machines to 27,000 primary schools.

Meanwhile, in February 1982, the BBC with government backing began its Computer Literacy project, based on an Acorn computer, a series of TV broadcasts and a book. A further £8 million was made available in 1983 for micros in schools and the training of teachers. The UK government also encouraged a highly successful chain of over 200 information technology centers (ITECs), providing hands-on experience in new technology for youths, especially unemployed youths in the cities. In 1984, £2.5 million was given to British School Technology, an independent national education center, to promote the teaching of new technology in schools and colleges. A further £5.5 million worth of schemes was announced in 1985 for the period 1986–8.

The French government, in a typically grand gesture, announced in 1985 a $250 million plan to put upwards of 120,000 micros in French schools as part of a massive crash program to improve computer literacy. The man put in charge by the French prime minister, Laurent Fabius, was Gilbert Trigano, the computer fan who had built up Club Med. A battle subsequently ensued between the nationalized French companies, Bull and Thomson, and Apple for a slice of the huge order. Bull was already linked with IBM via its IBM-compatible Micral series of personal computers.

Also in France could be found the Centre Mondial, a walk-in technology center off the Champs-Elysees in Paris, set up by author–politician Jacques Servan-Schreiber. The center, with its emphasis on open learning and the needs of the Thrid World, was inspired by the ideas of the Swiss educational psychologist Jean Piaget and South African-born MIT Professor, Seymour Papert.

Papert is associated with LOGO, a computer language derived from LISP and developed by the artificial intelligence community, which he argues is especially suitable for teaching children about computers. In his book, *Mindstorms: Children, Computers and Powerful Ideas* (Basic Books, New York, 1980), Papert says that LOGO is a much simpler and a more natural language to learn and requires no previous knowledge of mathematics or computers. He argues that experimenting in LOGO with a "turtle" – a hemispherical drawing device about 10 cm in diameter – gives children a sense of power over machines and develops intellectual structures in the mind that are based on mathematical concepts. Learning to communicate with computers thus becomes a "natural" process, he says, and learning to use computers can change the way children learn everything else.

Personal computers are also invading university and college campuses. Manufacturers are offering the same sorts of discounts and special deals as offered to schools in a bid to develop brand loyalty among the decision-makers of the future. Some colleges, such as Drexel University in Philadelphia, already specify ownership of a personal computer as an entry requirement, while others, such as Brown University, in Providence, Rhode Island, Carnegie-Mellon University in Pittsburgh and Stanford University in California, boast extensive networks of terminals and student workstations. The main problem – as in schools – is making good use of all the hardware now available and integrating the new systems into the college curriculum. Social scientists and college administrators are only just beginning to analyze their impact on social relations, library usage, scholarly habits and the learning process.

But are chips good for our children's health? Do schoolchildren and students really benefit from the use of personal computers? Obviously, just dumping computers on schools and colleges is not enough: in order to be useful, the new machines require decent software and trained teachers. Much of the software available in the early days of the personal computer invasion was very poor. Even today, much "educational" software is no more than glorified video games of limited value. Some simple drill-and-practice packages contain material that may just as well *not* be delivered by computer.

Critics like David Peck, Professor of Education at San Francisco State University, go further, arguing that the millions of dollars being spent on computers in schools might be better spent on more teachers, counsellors and textbooks. Schools, he says, should get "back to basics" and not waste money on gimmicks. Richard Clark, Professor of Educational Psychology at the University of Southern California, Los Angeles, says that computers have a "novelty effect" when introduced, but it soon wears off. He found that computer-instructed groups improved their scores for a while, because pupils invested more time and effort. But the effect was only temporary: once the novelty value disappears – as it did with TV-based and "audio-visual" instruction a decade before – the new medium seems to have no advantage, he says.

Joe Weizenbaum, MIT's Professor of Computer Science, has challenged the whole notion of "computer literacy" as "pure baloney." "People who use a computer only for the applications never need to learn how the technology works," he says. He is also concerned about the phenomenon of computer "nerds" or junkies – kids who become addicted to their machines and have little time for fellow humans. These unfeeling morons, he says, come to think it's possible "to reduce a human relationship to a print-out or to solve a moral question by bits and bytes" (*Business Week*, October 29, 1984 and *Time*, May 3, 1982).

Others have pointed to the unequal distribution of computers between schools and between the sexes. A 1982 US survey found that 80 percent of the 2000 most affluent schools had at least one computer, while 60 percent of the poorest had none. What's more, in poorer school districts it has been recently found that computers tend to be used for routine drill-and-practice, while it is the middle-class kids who are being taught to use them properly – thus increasing rather then reducing their head-start. As for unequal usage by the sexes, US and UK research published in 1986 found that boys outnumbered girls in computer classes by as much as three to one.

On the plus side, there is growing evidence that, in the 1980s, computers are being more successfully integrated into schools than they were in the 1970s. Computers are also of great help to the handicapped and have an important role to play as a remedial tool for slow pupils and those with other learning difficulties. There are many documented cases of computers bringing out talents in children not previously recognized. Even mathematics failures, perpetual truants and kids otherwise "written-off" by the school system have been unexpectedly turned on by computers and in a few cases have become programming prodigies. Computing expertise thus can act as a great leveller, and for some individuals micros can mean upward mobility.

6 Factories of the Future

Here Comes the Automatic Factory – The Robots Have Arrived – Computer-Aided Design and Manufacturing – Flexible Manufacturing Systems – Japan's Productivity Challenge to America and Europe – Factory Automation: Vision and Reality

Detroit, Michigan, 1986: A row of silver car bodies glides along the production line and halts. Six welders on either side immediately spring into action, moving swiftly around, over and inside the metal skeleton, sparking the many spot welds needed to hold the incipient automobile together. Within 23 seconds flat, they have knocked off 250 welds between them; the convoy moves on, and the next half-dozen cars are ready to be born. The welders never take a break, they don't drink coffee and they've never felt the need for conversation. No graffiti, no half-eaten sandwiches or soiled copies of the *National Enquirer* clutter their pristine workspace. That's because these ideal workers are robots, not people.

Nagoya, Japan, year 2001: From a group of small factories resembling a row of lock-up garages comes the faint hum of machinery. It is 3 a.m. and outside the streets are deserted. Inside, the factory floor is also deserted – there are no workers to be seen. Yet the assembly line is turning out finished goods as if it were the day shift. Robot vehicles move silently from machine to machine, carrying materials and removing machined parts. Automated pallets glide around the warehouse while more robots toil in the packaging department. Pre-programmed computers in the control room whir and blink as they process orders, draw on the data bank of computerized designs and convey a fresh set of instructions to the eerie, unmanned production area. The only humans to enter the factory will arrive to check things over in the morning.

Here Comes the Automatic Factory

The face of modern manufacturing industry is being changed out of all recognition by cheap computing power. New techniques of automated production are bringing about a revolution in manufacturing greater than anything since the Industrial Revolution. Many of the concepts are not new – the idea of "automation" itself has been around for years. What is making the difference today is the widespread availability of cheap, reliable and sophisticated control systems based on microelectronics. This – together with developments in numerical control, which make machine tools and robots more flexible and easy-to-use – means that highly complex automated factories can now be built at reasonable cost. The "Factory of the Future" is on the way.

There are basically two kinds of manufacturing industry. The *process control* industries like chemicals, plastics, steel and rubber produce materials, often in liquid or extruded form, through a continuous process largely controlled by switches, pressure valves and temperature gauges. Back in the 1950s and 1960s – in what has been called the "First Age of Automation" – it was the process control industries that were the first to be transformed because they were relatively simple, high-volume operations more amenable to automation.

The second type of manufacturing industry involves making physical changes to solids – most commonly metals, but also plastics, wood and composites – with the aid of machine tools. Thus the traditional metal-bashing industries have employed millions engaged in manufacturing parts by bending, battering, cutting, shaping and drilling bits of metal and then welding or assembling them together into finished products.

The new technologies make it possible to turn this type of manufacturing into an automated process, even if the production batches are quite small. Thus the ideal Factory of the Future would be run from the order book: orders and materials would go in one end, and would pass through the design, engineering, production and packaging stages and come out the other end, without being touched by human hand – or even being committed to paper. You've heard of the "paperless office": the Factory of the Future will also be the "paperless factory."

Machine tools date back to 1797, the year that Englishman Henry Maudslay designed the first screw-cutting lathe. Milling machines soon followed, and both types of machine changed little during the course of the nineteenth century. In fact, it was not until the 1950s that the first major innovation in machine tools occurred with the introduction of *numerically controlled* (NC) systems.

NC machines were run on a program consisting of holes punched in a paper tape which contained the instructions for the movement of the drill head or cutting tool. Their logic was thus "hard-wired" and inflexible, making it difficult to change the instructions. The late 1950s saw the introduction of *machining centers*, more sophisticated machines which can (usually) change their own tools and automatically carry out different tasks, such as drilling or turning.

The next big development was the arrival of *computer numerical control* (CNC) machines in the early 1960s. The addition of computer power meant that each machine could now store data and designs in its memory, making it possible to produce a range of goods on the one machine. The computer itself controls the operation of the machine tool and enables the human operator to carry out calculations and make adjustments much easier and quicker.

In the late 1960s, the Japanese started to link sets of NC or CNC machines to central computers which control their operations. This is called *direct numerical control* (DNC), and it has the advantage of dramatically reducing manpower, because only one human operator is required to control maybe a dozen machines in the linked production "cell," to use the jargon.

Robots have been talked about and tinkered with ever since Czech writer Karel Čapek's 1921 play, *Rossum's Universal Robots*, which brought the word into the English language. But it was the advance of the microchip in the 1970s that made possible enormous advances in robotics. The first generation of computer-controlled robots in the 1970s were called universal transfer devices (UTDs) and were little more than mechanical arms. They were also "deaf, dumb and blind" and very inflexible. They were used for such tasks as spot-welding, paint-spraying, loading and stacking.

But now, improvements in computer software and the development of vision systems ("robots that see") are transforming the prospects for these "steel collar workers." Robots are being used increasingly in general manufacturing, especially in sectors like the automobile industry. Robots enjoyed something of a boom in the early 1980s in the USA and Europe, as firms flocked to join the business. In Japan, Fujitsu Fanuc already has a plant where robots are helping make other robots. In another Japanese factory a robot recently crushed a human worker to death, causing the world's first high-tech murder.

Japan is also well ahead with "flexible manufacturing systems" (FMS). In this type of system a set of machines is linked up by means of handling devices in such a way that parts can be passed automatically from one to another for different manufacturing processes. Even more important, FMS (or "advanced manufacturing technologies" (AMT), as they are sometimes called) enable a company to produce small batches of components or vary the product specifications as quickly and efficiently as if it were using a mass production line dedicated to one product. A central computer controls the whole process and keeps a check on the whereabouts of each part.

Another hot new area undergoing a boom is computer-aided design (CAD), in which draftsmen's drawings can be created on a VDT screen and then be manipulated, updated and stored on a computer without even the use of a pencil. CAD leads to CAE (computer-aided engineering), in which the three-dimensional designs can be examined, analyzed and even "tested" to simulated destruction without ever having been made!

CAE, in turn, leads to CAM (computer-aided manufacturing), because the same design and prototype testing process generates a common database about the product from which a set of instructions can be derived for actually manufacturing it. The final stage, computer integrated manufacturing (or CIM, pronounced "Sim"), will see the linking up of production cells and "islands" of automation into one comprehensive, integrated manufacturing system controlled by computers.

CIM was very much the buzzword concept of the early 1980s and really amounts to a bringing up to date of the early 1950s idea of what "automation" and the "automated factory" meant. It has taken 30 years of research and development – plus the invention of the microchip – to make the vision of the completely unmanned, automated factory a distinct possibility. As the sophisticated software needed to "glue" the whole array of hardware together – in particular, the manufacturing automation protocol (MAP) – is further developed, it may be only a few years before that vision becomes a reality.

The Robots Have Arrived

In Čapek's 1921 play, hero engineer Rossum creates a new breed of robots to do the world's dirty work. He took the name for his human-like creatures from the Czech word *robota*, meaning forced or slave labor. After taking on all the humans' unpleasant jobs and fighting their wars, Rossum's robots eventually rise up and take over the world, like so many Frankenstein's monsters.

Unfortunately, this has given "robots" a bad name and an unjustified reputation, which has been perpetuated down through the years – for example, by Isaac Asimov's science fiction writings of the 1940s and films like *Star Wars*, which featured the robot, R2-D2. Today's industrial robots are really nothing like Čapek's humanoids. They cannot see, hear, smell or think. They are simply machine tools, typically no more than a mechanical arm controlled by a computer which can be programmed to carry out different movements. In fact, the Robot Institute of America offers the following definition of a robot: "A reprogrammable, multifunctional manipulator designed to move material, parts, tools or specialized devices, through variable programmed motions for the performance of a variety of tasks."

The age of the industrial robot really began in 1946, when US inventor George Devol developed a memory device for controlling machines. The first patent for a programmable arm was filed by Devol in 1954 and in 1960 Devol teamed-up with the charismatic American engineer Joe Engelberger to form Unimation, the first company to make and sell industrial robots. By using the word "robot" to generate interest in their product, they actually succeeded in confusing everyone!

But the marketing ploy worked in that Unimation's "robots" generated piles of press cuttings – if not revenue. In fact, industrial robot sales were very slow to take off: the world's first industrial robot was installed in 1961 at a General Motors factory in New Jersey, but the high costs and teething troubles delayed subsequent installations to such an extent that Unimation failed to show a profit until 1975. But the *potential* of robots to greatly increase productivity and reduce manpower was steadily growing more apparent.

Peter B. Scott points out, in his book, *The Robotics Revolution* (Basil Blackwell, Oxford and New York, 1984), that these so-called "first-generation" robots were ideal for dirty and repetitive jobs such as welding, paint-spraying, grinding, molding and casting. Of the 4700 robots in use in the USA in 1981, for instance, 1500 were used for welding, 850 in foundries, 840 for loading and 540 for paint-spraying. Only 100 were used for assembly tasks. Scott's "second-generation" machines are able to "see" and "touch," while the coming "third generation" will have intelligence and massive computing power, enabling them, for example, to "infer logically" – that is, to work out for themselves how to do something.

Three key developments sparked off the robotics boom in the early 1980s. The first, of course, was the arrival of the microchip, which meant that the computer "brains" of robots were much cheaper and took up less space. Second, the fear of foreign competition – particularly from the Japanese – and the revelation that US productivity gains in the 1970s were miniscule and were lagging behind the rest of the world stirred US manufacturers into action. For instance, between 1947 and 1965 US manufacturing productivity grew by 3.4 percent a year on average, but this dropped to 2.3 percent in 1965–75 and below 1 percent in the late 1970s. Third, wage inflation and extra fringe benefits were dramatically increasing the costs of human assembly-line workers relative to robots. The incentive to invest in new technology was greater.

Furthermore, a number of systems up-and-running – particularly in the automobile industry – had clearly demonstrated the productivity gains to be obtained from using robots in certain situations. Workforce representatives were generally impressed by their successful application in hazardous environments such as furnaces, chemical factories, nuclear power stations, in outer space and on the ocean floor. Besides, there was no great merit in having people work in dirty, noisy, dusty and dangerous jobs.

By the end of 1984, according to British Robot Association figures, there were about 100,000 robot devices at work in the world, of which 64,600 were in Japan, 13,000 in the USA, 6600 in West Germany, 3380 in France, 2850 in Italy and 2600 in the UK. On a per capita basis, the European order was Sweden first, with about 19 robots per 10,000 production workers, followed by Belgium and West Germany, with only 3 per 10,000 workers, about the same as the USA and far less than

Japan. By the end of 1985, the USA had 20,000, West Germany 8800 and the UK 3200.

In the mad scramble to get into robotics in the USA in the early 1980s, computer giant IBM entered the market with two home-grown robots, while General Electric (partly through its acquisition of Calma) and Westinghouse (which bought Unimation in 1982) aimed to offer "complete solutions" to factory automation, making available a whole range of equipment. But some commentators alleged that the robotics revolution had been overblown and there were now too many players (perhaps 250 worldwide) competing in what was still a fairly small market in relative terms. As Joe Engelberger once quipped, "I sold the idea of robots to my competitors, not my customers." A shakeout of some kind seems inevitable: the suggestion is that many of the smaller, specialist robot companies will be forced out of business or into merger with larger companies which have the resources needed to compete in the Factory-of-the-Future market. For example, American Robot Corporation sold out to Ford in 1985.

Not every firm's experience with robots has been exactly trouble-free, however, and customers have become more cautious in recent years. One major factor holding back the robot revolution has been management resistance to change. Managers either don't want to know about new methods or fear a hostile reaction from the shop floor. Foremen and supervisors worry that their authority will be undermined or they will be out of a job. Despite the glamor and the hoopla surrounding robots, there are many companies in the USA and Europe that are so conservative that they haven't even investigated the possibility of using robots. Most of them will eventually fail.

There have also been widespread fears of labor displacement. The American Society of Manufacturing Engineers predicts that many thousands of jobs will go by 1990. In their book, *Robotics: Applications and Social Implications* (Ballinger, Cambridge, Mass., 1983), Robert U. Ayres and Steven M. Miller of Carnegie-Mellon University in Pittsburgh point out that over half the jobs that could be replaced by robots are in the five metalworking sectors and almost half the workers in those five sectors are located in plants in the five Great Lakes states: Indiana, Illinois, Michigan, Ohio and Wisconsin.

Robots will have a disproportionate impact on racial minorities, women and members of labor unions, and the age structure of manufacturing industry is such that it will be impossible to avoid redundancies through attrition. Industry and government, say Ayres and Miller, should plan for the changes ahead by putting more money into education, retraining in new skills and relocation schemes to ease the transition to the Factory of the Future. Some union leaders are more skeptical and ask whether there will be any skilled jobs for the workers to be retrained for.

The next generation of robots will be "smart" in that they will have greatly improved vision and sense of touch. Work on developing "vision systems," or machines that "see," began in the late 1970s at Unimation, Westingthouse, Texas Instruments and IBM, but it is currently GM that is pushing forward the technological frontiers with its purchase, in August 1984, of major stakes in four small vision system companies: Automatix, Robotic Vision Systems, View Engineering and Diffracto Ltd. In fact, in 1985–6 machine vision became the hottest of all the technologies of the Factory-of-the-Future scenario. This is because cheap processing power and new software now make it possible for digital computers to analyze and interpret images portrayed as a matrix of dots.

Thus, machine vision offers the prospect of tireless and totally reliable robot eyes on the production line which can replace, for example, human quality controllers who all too often possess human fallibility. Vision systems will "close the loop" on automated process control with visual feedback mechanisms: if a part is defective, it will be automatically rejected. When the sensors, computers, robots and machine tools are all tied together – that is, when the so-called "compatibility" problem is solved by the adoption of a standard like GM's manufacturing automation protocol (MAP) – we will have taken another giant step toward the factory of the future.

The problem at present is that the new vision systems are expensive and are slow in getting out of the lab and onto the factory floor because of management and workforce skepticism. But with GM and Ford – as well as dynamic specialist companies like Perceptron Inc. and Machine Vision International – behind the push to get artificial vision installed in factories across the land, the prospects for robots that "see" are brighter.

The future of robotics lies with machines that "think." Again, it is GM and other automobile makers like Ford who are pushing things forward, following GM's joint venture with Fanuc ("GMFanuc") and its investment in Teknowledge, an AI company in Palo Alto, California. Of course, robots that exhibit real "intelligence" are years away, and their prospects depend heavily on developments in AI research. But robots that not only see, feel and hear but also "think" in certain respects are a definite possibility for the 1990s.

Computer-Aided Design and Manufacturing

Of central importance to the Factory-of-the-Future concept are developments in computer-aided design and manufacturing (CAD/CAM). Between the design and manufacturing stages comes the test stage, computer-aided engineering (CAE); and, taken together, CAD/CAE/CAM systems represent a much larger market than either robotics or flexible manufacturing systems. As they move downmarket in the late 1980s with the spread of the IBM PC, sales of CAD/CAM systems are growing by around 30 percent a year. Out of today's CAD/CAM systems will develop the computer-integrated manufacturing (CIM) systems of tomorrow.

The CAD part of CAD/CAM has its roots in the early 1960s when computerized draftsmen's aids were first displayed. For example, the SKETCHPAD system developed at MIT in 1963 used a light pen wired to a large (and very expensive) computer. One-time industry leader, Computervision Corporation, launched its first CAD/CAM system, in 1969. The arrival of the microchip in the 1970s transformed the CAD business, with the result that great strides have been made in CAD systems in the last decade or so.

CAD enables a designer or draftsman to construct detailed drawings of machinery, parts or circuits without ever touching a pencil, ruler or compass. Instead, the designer uses a screen and keyboard, moving a special stylus over the screen or a separate graphics tablet alongside. Standard shapes and shading can be simply summoned up from the computer's memory and added to the "drawing." With CAD, designers can turn their drawings around, enlarge them, color them or slice them in half on the screen in order to examine or modify them. If they wish to pore over their drawings, a paper copy can be provided at the touch of a button.

From the original "electronic pencils," CAD systems today have developed to a point where they offer the facilities of three-dimensional or solid modelling, engineering analysis and testing, simulation and interfacing with machine tools, as well as automated drafting. This enables manufacturers to design, to build life-like models and to test, for example, projected automobiles, airplanes and oil drilling platforms – all on the computer screen. Once the geometric model has been completely defined, instructions for manufacturing it – even down to the tool-cutting paths – can be derived from the system's database. CAD, CAE and CAM have therefore effectively merged into one, although most people use the term "CAD/CAM."

CAD/CAM has rapidly established itself in the automobile and aerospace industries. Companies like Chrysler, GM, Boeing and Pratt & Whitney are major users in the USA, while Ford and Austin Rover are pioneering its use in Britain. CAD/CAM not only allows designers to see what a car or plane will look like when built, but it can also be used to see how they will behave under certain in-service conditions. The whole new field of computational fluid dynamics (CFD) has opened up. Finite element analysis, the mathematical method of calculating the stresses on, for example, an airplane's wing or an automobile bumper, allow realistic computer simulations to be carried out, which show exactly how the product will perform in real life. Even slow motion "movies" of the action can be made. Again, plummeting computer costs now make the necessary complex computer calculations much easier to carry out.

CAD/CAM is especially useful in routing. Airplanes and oil rigs contain many miles of pipes, and working out the best way to route them can be a designer's nightmare. Now CAD/CAM systems can locate the best possible paths in a matter of seconds. McDonnell Douglas, for example, has a system that not only plots optimum paths for the three miles of hydraulic tubing that twist and turn their way through a DC-10's airframe, but also issues instructions on how to manufacture them from standard parts. Bechtel, the giant plant construction company, is spending millions of dollars on CAD/CAM systems for plotting pipework.

Why stop at large objects? CAD/CAM is widely used in the electronics industry to design and test tiny microchips. New systems can work out the optimum layout for integrated circuits and the best route for each connection. Another CAD/CAM function is "modal testing," whereby the designer can pinpoint the moving part in a machine which is or will be responsible for a particular vibration. CAD/CAM is also being used for management systems and process planning, that is, designing and simulating the operations of an entire factory, warehouse or supermarket before it is built. In this way, potential bottlenecks and other problems can be identified and ironed out well in advance.

CAD/CAM thus has many advantages for users. It can greatly increase the productivity of expensive design staff, eliminate boring and repetitive tasks, cut the lead-time from design stage to final product, reduce design and manufacturing errors, increase product quality, cut the cost of updating or modifying products, and enable orders to be repeated with little or no delay. In short, CAD/CAM is a company's own route to greater competitiveness in world markets.

There are obstacles, however, to its implementation. One is the great diversity of hardware and software available and the lack of compatible standards. Another is the very complexity of the manufacturing operations of many companies, which are not always amenable to computerization and have evolved in different ways to other firms over the years. In some companies, design and manufacturing departments are still segregated and it is often difficult to get them working together. Management and union conservatism have not helped.

In the older, less adaptable, nations of Europe, slow takeup is a problem: in 1982, for example, the British government allocated over £12 million to make British companies more aware of CAD/CAM and to encourage its use. But of the 2000 or so firms receiving grants for advice and consultancy on CAD/CAM, fewer than 250 took the matter further, and even fewer actually got round to installing CAD/CAM systems.

Top companies in the CAD market in the USA are IBM (21 percent of sales in 1985), Intergraph (15 percent) and 1984 market leader Computervision (12 percent) followed by Calma (a disappointment for owner GE), McAuto (McDonnell Douglas Automation), Daisy Systems, Mentor Graphics, Prime Computer, Control Data and Applicon (Schlumberger). Apollo Computer and Sun Microsystems are doing well with their engineering workstations.

Autodesk of California is doing a lot of business with its AutoCAD software for the IBM PC. This enables engineers, architects and scientists to do complex CAD work which would have required an expensive minicomputer only three or four years ago. Now they can use a PC to suggest designs to clients or to plan the floor layouts in new office blocks. New simulation techniques even allow architects and the clients to "walk around" and "through" new buildings before they are built. Desktop systems for scientists to simulate experiments are also catching on fast.

In CAE, on the other hand, Daisy, Valid Logic and Mentor are still holding their own, although they are increasingly threatened by the IBM PC derivatives. In 1986, for example, IBM introduced its RT PC, a new type of high-performance CAD/CAM workstation which uses reduced instruction set computer (RISC) architecture.

The prospects for CIM depend in part on continued developments in CAD/CAE/CAM such as solid modelling, increased workstation power and the speeding of data transfer between workstations. But most important of all, CIM requires the adoption of common standards or protocols to link up the various "islands" of automation supplied by different manufacturers. Leading the way are GM's MAP (manufacturing automation protocol) and Boeing's TOP (technical office protocol), which were demonstrated for the first time at the Detroit "Autofact '85" show. MAP and TOP embrace OSI (Open System Interconnect), which is now being pushed by the Washington-based International Standards Organization (ISO).

Manufacturers such as Deere, Dupont, Ford, Kodak and McDonnell-Douglas have followed GM and Boeing's lead, while IBM, AT&T, DEC and GE are also committing themselves to MAP and TOP. In 1986 GM opened a truck assembly plant at Pontiac, Michigan, which uses MAP chips, and Japanese robot maker Fanuc announced that it was also adopting MAP. Allen-Bradley, a manufacturer of industrial controls based in Milwaukee, opened what is probably the best example of a CIM facility in the world, which makes industrial starters and controllers on a flexible, flowing assembly line; the firm says the alternatives were either to go offshore or to go out of business.

But as the recent Office of Technology Assessment report (*Computerized Manufacturing Automation: Employment, Education and the Workplace*, US Congress, Washington DC, 1984) pointed out, there are still many socioeconomic barriers to be overcome before we see the widespread introduction of CIM, such as the high cost and resistance on the factory floor and in the boardroom. The CIM market is newly emerging, but how quickly it will emerge also depends crucially on developments in flexible manufacturing systems.

Flexible Manufacturing Systems

Way back in 1965, a British producer of cigarette-making equipment called Molins brought out a revolutionary machine tool called System 24. Designed by Theo Williamson, System 24 was an advanced automated machine capable of running for 24 hours a day and of making small batches as efficiently as long production runs. Unfortunately, Williamson was 15 years ahead of his time: although it was sold to Rolls Royce, IBM and Texas Instruments, System 24 absorbed huge sums of development money, and by 1979 Molins plunged into the red. In 1973 the company was forced to close down its entire machine tool division.

The timing could not have been more tragic. For at the very moment that Molins was folding, cheap microelectronics was enabling US firms like Cincinnati Milacron, White Consolidated and Kearney & Trecker to develop less costly flexible manufacturing systems to replace the older "hard" automation systems of the 1950s and 1960s. The Japanese followed in 1977 and soon took the technological lead.

"Hard" automation means that a production line is dedicated to turning out one product in huge volume. Flexible manufacturing systems (FMS) enable a company to produce goods in small volumes as cheaply and efficiently as if it were using mass production methods. Since most manufacturing these days is in small "batches" or variants of similar products for different markets, the advantages of flexibility are obvious.

In fact, the savings and productivity gains made possible by FMS can be phenomenal, because a single, all-embracing system can replace several conventional machining lines. FMS yields enormous savings on labor – most assembly work in factories is still labor-intensive. It also saves on plant space and further decreases capital requirements by dramatically reducing the amount of work-in-progress and stock inventory. Product quality is better, so wastage and rectification costs are reduced. Plant utilization is improved because the machines can be kept running most of the time and re-programmed quickly. Reduced lead-times also enable the manufacturer to react much more swiftly to market trends. Flexible manufacturing systems therefore represent not just a new technology, but an entirely new way of thinking.

Economists refer to these advantages of flexible automation as "economies of scope," as opposed to "economies of scale." Under old-fashioned "hard" automation or mass-production, the greatest savings were realized only with large-scale plants and long production runs. The FMS revolution makes similar economies possible at a wide range of scales. Economies of scope break all the rules of traditional manufacturing. There is no long trip down the learning curve; and entrepreneurial newcomers from unheard-of places can enter a field as never before and swiftly overcome less agile older producers or nations. This is the reality of international competition in the age of FMS.

Of the 120 or so "true" FMS installations in the world, about half of them are located in Japan, where the major suppliers are Yamazaki, Hitachi-Seiki and Fujitsu Fanuc. With their long-run view of manufacturing, their just-in-time concept of parts supply and their emphasis on preventative maintenance, the Japanese have taken to FMS like ducks to water.

At Fujitsu Fanuc's futuristic factory near Mount Fuji, 30 manufacturing cells and just 100 workers turn out as many robot parts as would be produced in a conventional plant costing ten times as much and employing ten times as many people. In Fanuc's nearby electric motor factory, 60 cells and 101 robots toil night and day automatically turning out 10,000 electric motors a month.

Yamazaki's $20 million flexible automation plant near Nagoya employs just 12 workers in the day and one nightwatchman. Its conventional equivalent, says Yamazaki, would require 215 workers and nearly four times as many machines, and would take three months to turn out the machine tool parts the new plant makes in three days.

Another "twenty-first-century" Yamazaki plant 20 miles away is run by telephone from company headquarters and employs about 200 workers instead of 2500 in a conventional factory. It can cope with sales varying from $80 to $230 million a year without having to lay off workers.

US companies, on the other hand, have been slower to install FMS, despite the availability of advanced systems from leading machine tool firms like White, Ingersoll, Kearney & Trecker and Cincinnati Milacron. Critics say that US managers, unlike their Japanese counterparts, lack the technical expertise to understand the complex new systems and are too worried about the short term to think long term. Recent research by Professor Ramchandran Jaikumar of the Harvard Business School found that US managers were using FMS far less flexibly than the Japanese or West Germans. And they were altogether too hooked on the idea of keeping old machine tools going: according to one survey, 34 percent of US machine tools in use are over 20 years old and only 4 percent are numerically controlled – that's an even worse record than Britain's!

Deere & Co, however, are one exception, with their giant new tractor assembly plant at Waterloo, Iowa, which boosted the competitiveness of its products (although lack of demand and a labor dispute have since caused a shutdown). General Electric has transformed the prospects of some of its older plants (like its meter factory in New Hampshire and its locomotive factory at Erie, Pennsylvania) through flexible automation – making them what it calls "factories *with* a future." The British government also has a program to boost the use of flexible technologies in traditional industries like clothing, textiles and footwear.

But again, it is the automobile industry that has made the greatest strides. GM and Ford, for example, are pushing ahead with FMS installations in a desperate effort to stay competitive with the Japanese. In their book, *The Future of the Automobile* (MIT Press, Cambridge, Mass., 1984), Alan Altshuler, Martin Anderson, Daniel Jones, Daniel Roos and James Womack argue that the automobile industry is undergoing its "fourth transformation" as high technology in the form of FMS is applied to the manufacture of cars. (The first three were Model "T"-style mass-production, postwar European innovations in car design like the Volkswagen Beetle and the Japanese quality assault in the 1970s.)

Flexible automation, say these authors, will have even greater consequences for the auto industry than earlier transformations. With CAD/CAM, FMS will allow manufacturers to move from design concept to prototype to production far more quickly; it will revolutionize actual production methods and will enable car makers to target a far greater variety of models to specialist niche markets. Until recently, most observers saw the major trend in the world automobile market as being toward a small, standardized car. Now FMS will reverse that process, enabling producers to develop cars for specific markets without incurring the traditional costs of small-scale batch production. And because FMS will reduce the need for labor, the expected shift to low-wage countries will not now occur on a large scale. In fact, car plants will be able to be located much nearer the markets where the cars are sold.

Despite encouragement by governments, takeup of FMS in Europe has been typically slow, but there are some encouraging signs. In Britain, Cincinnati Milacron (UK) and Kearney & Trecker Marwin (UK) have showpiece systems at Birmingham and Brighton, respectively. IBM uses FMS extensively in its computer plant at Greenock, Scotland. Construction materials company Travis & Arnold has a new flexible automated wood-cutting mill at King's Lynn, Norfolk – the first such change in saw mill technology for over 20 years. Aerospace company Normalair–Garrett has an FMS facility at Crewkerne, Somerset, which has notched up some pretty amazing productivity gains: labor costs are down from £400,000 to £150,000 a year, output per worker has risen from £70,000 to £210,000 a year and work-in-progress is turning over 24 times a year instead of just 3.3 times – with a magical reduction in the value held at any one time from £690,000 to only £90,000.

That's why FMS can pay for itself so quickly. Rolls Royce, the aero-engine maker, has managed to finance a massive investment in FMS at its Derby, England, works solely on savings in working capital. With some parts for Boeing 757 engines costing over £10,000 apiece and lead-times exceptionally long, the incentive to reduce work-in-progress is obvious. Now, 7 robot cells making turbine blades employ 6 men instead of 30. Each blade used to take 6 minutes to make: now it takes 45 seconds. Overall productivity is up about 30 percent. A final British example: at its Colchester, Essex, plant, the 600 Group's FMS factory can produce machine parts from the order book in three days rather than three months and employs 3 men rather than 30.

Elsewhere in Europe, major users of FMS are Volvo in Sweden and Renault in France. In West Germany, Messerschmitt–Bölkow–Blohm's plant in Augsburg has achieved impressive savings in the machining of parts for the Tornado jet fighter. Machine tools are run for an incredible 75 percent of the time, and the total production time, the number of machines and the number of workers needed have all been halved. Overall annual costs have been cut by a quarter.

Flexible manufacturing systems do have some disadvantages, however. They are of course very expensive, they are difficult to install, and they do not work properly unless there are reliable linked systems – for example, for materials handling and automated storage and retrieval. Manufacturers looking for a "quick fix" technological solution to their problems simply by purchasing a "turnkey" system (that is, a complete, ready-to-go system – all you have to do is turn the key) will be disappointed.

Some say that the technology of FMS has been oversold. FMS installations, like anything else, have to be worked at and constantly improved by people who know what they're doing. Moveover, unless the correct management and marketing decisions are taken, no amount of new hardware can stop a sick company from losing market share. FMS must be integrated with the business as a whole; otherwise its great benefits will be dissipated.

Japan's Productivity Challenge to America and Europe

At the end of World War II, US manufacturing industry reigned supreme. Between 1940 and 1945, Detroit's factories had churned out 50,000 tanks, 600,000 trucks and 4 million engines. In five years, America's war machine also produced an astonishing 5200 ships, 300,000 warplanes and 6 million guns. In fact, it was largely US industrial might that eventually won the war on the Western front for the Allies. As "Big Bill" Knudsen, GM's president who headed the wartime industrial effort, once commented: "We smothered the enemy in an avalanche of production."

But success in war bred overconfidence in peacetime, which bred complacency. In the 1950s and early 1960s, it was felt that US industry could do no wrong. The techniques of mass-production had been conquered: there was nothing left to learn. On top of that, industry was not attracting the right kind of people. As in Britain and Europe, the best and the brightest tended to steer clear of manufacturing, preferring to go into better-paid and supposedly more glamorous fields like law, medicine, the mass media and politics.

Then, in the 1970s, the Japanese tidal wave hit. A flood of manufactured imports – especially consumer goods such as motorcycles, automobiles, TVs, stereos and cameras – poured into America. The Japanese invaders were all-conquering, offering lower prices because of higher manufacturing productivity and setting standards in product quality never achieved before with mass-produced goods. By skilfully "targeting" vulnerable areas and manufacturers, often with the advice and financial assistance of the giant Ministry of International Trade and Industry (MITI), the Japanese were picking off one market after another.

For example, it was obvious that the Japanese had targeted the machine tool industry, robotics and FMS as part of their own "Factory-of-the-Future" plan. The Japanese see machine tools as "the mother of machines" because the efficiency of machine tool design determines in large part the efficiency of the production process. With massive resources available to it, the comparatively young Japanese machine tool industry was by 1975 exporting 40 percent of its production, of which nearly two-thirds was numerically controlled. Japan swiftly overtook countries like the UK (whose Alfred Herbert Ltd had earlier helped the Japanese get started) and France to become No. 3 in machine tools behind the USA and West Germany in the late 1970s. By the early 1980s Japan had overtaken the USA to become No. 2 and by the end of 1986 had overtaken West Germany to become No. 1 worldwide. As we have seen, Japanese robotics and FMS have enjoyed an equally meteoric rise.

US manufacturers were taken by surprise. Starting (almost literally) from scratch in 1945, the Japanese in 30 years had developed their production techniques to a point where they were not only teaching the rest of us a thing or two, but were winning the productivity battle hands down, as Richard J. Schonberger explained in his book *Japanese Manufacturing Techniques* (Free Press, New York, 1982). Just as was said of West Germany, the Japanese had lost the war but were now winning the peace.

The story of the Japanese assault on world markets has been better told elsewhere and "learning from Japan" became a major growth industry in itself. Management consultants quickly swung into action, exhorting US employers to start Japanese-style "quality circles." Fear of Japan produced a huge literature on the "secrets" of Japanese success and what the USA should do about it, most notably the books of William Ouchi, *Theory Z: How American Business Can Meet the Japanese Challenge* (Addison-Wesley, Reading, Mass., 1981); Richard Tanner Pascale and Anthony G. Athos, *The Art of Japanese Management: Implications for American Executives* (Simon and Schuster, New York, 1981); and Chalmers Johnson, *MITI and the Japanese Miracle* (Stanford University Press, 1982). Other texts, such as Jared Taylor's *Shadows of the Rising Sun* (William Morrow, New York, 1984), sought to demystify the Japanese "miracle."

By 1980 US manufacturers and the federal government were in a veritable state of panic. New productivity figures showed clearly that the growth in US manufacturing productivity had fallen from 3.4 percent in the 1950s and early 1960s to less than 1 percent in the late 1970s. There were widespread fears that US industry could no longer compete. US companies were undergoing a "hollowing out" as more and more became assembly plants for imported parts. In Detroit, Chrysler went bust, just as the Japanese grabbed 27 percent of the domestic automobile market. A famous article by Harvard University's Robert Hayes and William J. Abernathy in the *Harvard Business Review* ("Managing Our Way to Economic Decline") and a landmark *Business Week* special issue, "The Reindustrialization of America" (June 30, 1980) signalled that the USA was facing a kind of national productivity emergency.

Another flood of books and articles followed. In the newly created *National Productivity Review*, Robert U. Ayres and Steven M. Miller of Carnegie-Mellon University confirmed the US productivity decline and argued for more investment in the new technologies of robotics and CAD/CAM. William J. Abernathy, Kim B. Clark and Alan M. Kantrow came out with *Industrial Renaissance: Producing a Competitive Future for America* (Basic Books, New York, 1983), while Robert U. Ayres followed up with *The Next Industrial Revolution: Reviving Industry through Innovation* (Ballinger, Cambridge, Mass., 1984).

Robert B. Reich, also at Harvard, penned numerous articles calling for the US government to adopt a European-style industrial policy. In his book, *The Next American Frontier* (Times Books, New York, 1983), Reich argued for subsidies, tax breaks, generous R&D grants and a measure of enlightened protectionism to ease the transition from "Rust Bowl" to "Sunrise" industries. In addition, he called for social reorganization to replace competition with cooperation through greater participation and shared ownership. A similar plea is made by MIT's Michael J. Piore and Charles F. Sabel in their book, *The Second Industrial Divide: Possibilities for Prosperity* (Basic Books, New York, 1985). They argue for government planning of the switch to CIM in industrial countries.

But the case for industrial policy had already been undermined by the continuing US boom, US success at job creation and, most important of all, the news that US productivity was undergoing a remarkable turnaround. New figures revealed that the annual growth in US manufacturing productivity had gone back up to 2.7 percent in the period 1980–4 – approaching the levels of the 1950s and early 1960s, and more than twice the level of the late 1970s. The great productivity emergency seemed to be over.

Detroit's traumas had a lot to do with the turnaround. Automobile industry managers were shocked into changing their production methods – in some cases for the first time in years. Productivity was up 50 percent at Chrysler under Lee Iacocca, 30 percent at Ford and 14 percent at GM between 1980 and 1984. A good dose of old-fashioned competition converted management to the cause of productivity, and the reality of mass lay-offs worked wonders with labor. But other industries had equally rude awakenings: Xerox, for example, thought it had everything going for it until the Japanese in the shape of Minolta and Ricoh walked in and grabbed a significant slice of the copier market. Xerox was forced to respond with its own productivity and quality blitzes.

Bit by bit, US industry is preparing for the twenty-first century, as is evidenced by the Deere tractor factory at Waterloo, Iowa, General Electric's meter and locomotive plants, Apple's futuristic Macintosh facility at Fremont, California and AT&T's printed circuit board plant at Richmond, Virginia. People are beginning to understand the importance of manufacturing. A new wave of interest in manufacturing technology has also been spreading through US universities and colleges, which have previously preferred to concentrate on more esoteric matters. To today's graduates, "manufacturing" is no longer a dirty word.

But the Japanese are determined to stay one jump ahead, with ambitious projects like their advanced manufacturing technologies test-bed at Tsukuba Science City near Tokyo and their commercial devices such as collaborative agreements with foreign firms possessing know-how they wish to get their hands on. Some British firms, for example, have found that these supposedly joint projects tend to work in only one partner's favor.

The story of Britain's decline from being world leader in manufacturing obviously goes back a long way. Many theories have been advanced to account for it, ranging from the idea that Britain became "over-extended" with the burdens of Empire to the argument that the governing classes of the eighteenth and nineteenth centuries never really took to industry, generally preferring the life of country squires. Industry was seen literally as a blot on the landscape. More recent theories have highlighted the size of the "unproductive" public sector and unhelpful bank lending policies. Mancur Olson's thesis that stable societies suffer from "institutional sclerosis" in the form of cartels, special-interest lobbies and monopolies, outlined in his book, *The Rise and Decline of Nations* (Yale University Press, New Haven, Conn., 1982), would also seem to be particularly relevant to the British case. Special interests and class rigidities, Olson says, are harmful to economic growth, innovation and enterprise. As a result, productivity plunges and there is a collapse in competitiveness.

But Britain also underwent its own kind of productivity panic in 1979–80, just after the microchip became a hot news item. In May 1979 the Ingersoll Report slammed British managers for complacency and charged that Britain had squandered an early technological lead in robotics, just as it had done with FMS (not to mention computers, biotechnology and just about everything else). In November 1979 and February 1980, reports commissioned by the British Cabinet from the Advisory Council for Applied Research and Development (ACARD) on robotics and CAD/CAM delivered essentially the same message: Britain's industrial strength was ebbing away as the country fell further behind in the technology race. Urgent action was needed to help modernize its out-of-date factories.

Despite support from the National Economic Development Office, the UK government remained wary of state intervention. Prime Minister Thatcher argued that it was the job of British managers, not the government, to introduce new production techniques; there was not a lot that Whitehall could do. This was a view that gained some support when it was learned that only one British company, the 600 Group, had applied for government grants that were *already* available to assist in the start-up of FMS projects.

Even so, the British government obviously had second thoughts after the crisis in the UK machine tool industry became even more apparent in 1981. Alfred Herbert, once a world leader in machine tools, went out of business. A rash of articles and numerous conferences highlighted how British industry was lagging behind the rest of the world. A major study by the Policy Studies Institute (PSI) revealed that only 30 percent of British manufacturing firms were using microelectronics in their products or production processes – or even planning to do so!

In October 1981 the new Information Technology minister, Kenneth Baker, announced a government-backed CAD/CAM awareness campaign. This was followed in 1982 and 1983 with a £60 million FMS scheme, an extra £12 million for CAD/CAM awareness and a further £5 million for direct CAD/CAM investment support. By 1984 there were definite signs of an increase in capital investment in factory automation. A PSI follow-up survey, reported in March 1984, was also more optimistic, finding that the percentage of firms using or planning to use microelectronics had jumped from 30 to 47 percent in just two years; while in 1985 an international PSI study put Britain's record in a European context – just behind West Germany, but ahead of France. However, all the European nations surveyed were way behind Japan.

Factory Automation: Vision and Reality

Not everyone is totally enthusiastic about the Factory of the Future or wholly optimistic about its prospects. There is growing skepticism in certain quarters, both about the rate of progress toward CIM and about

the possible impact of automation on the quality of worklife. As we move (albeit slowly) out of the "showcase" or token automation phase to the more general application of new systems in manufacturing industry, the technical and managerial problems will become more acute – and the debate about the benefits of automation more intense.

Use of the words "automatic" and "automation" can in fact be traced back to 1586. "Automatic" was used widely in the nineteenth century, and even in the 1920s and 1930s writers were proclaiming the dawn of the "automatic factory" and the widespread use of robots. In the postwar period, Norbert Wiener's pessimistic *Cybernetics* (John Wiley, New York, 1948) and *The Human Use of Human Beings* (Houghton Mifflin, New York, 1950) and John Diebold's optimistic *Automation: The Advent of the Automatic Factory* (Van Nostrand, New York, 1952) sparked a wave of interest in automation and its social consequences. But, despite important studies like James R. Bright's *Automation and Management* (Harvard University Press, Cambridge, Mass., 1958), interest in the subject waned in the 1960s when it became apparent that progress toward the automated factory was painfully slow.

The arrival of information technology has opened up vast new possibilities, but even so it is important to be realistic about the pace of developments. Skeptics point out that robots are still very limited in the work they can do and that robots that resemble humans by being able to sense their environment are a long way off. CAD installations have not always been a success, and making the connections between CAD terminals and CAM machine tools has proved difficult and expensive. For example, in the aerospace industry – a heavy user of CAD/CAM – it still takes engineers more than ten years to design and build a new plane, and the failure to link CAD and CAM systems adequately is proving to be a major bottleneck.

Even NC machine tools, which have been around for decades, are not as widely used in industry as might be expected. In the USA, for example, only 2 percent of machine tools in the metalworking industries in 1978 were numerically controlled. By 1983 the figure had grown only to 4.7 percent. CNC and DNC machine tools are even rarer. At the other end of the scale, the number of FMS systems actually in operation worldwide is only just over a hundred, as we have noted. The very complexity of automating batch production has proved to be a major obstacle, and even those systems that are up-and-running are not very flexible. They also tend not to embrace storage and materials handling areas which have proved difficult to automate, despite the development of AGVs (automated guided vehicles). Although FMS systems have the greatest potential, critics say that their evident expense, complexity and essentially integrated nature (which means that a single component failure shuts down the entire system) have clearly put off many potential customers. There are few compelling "shop window" installations.

Linking up all the "islands" of automation into complete CIM systems may take much longer than expected, according to many observers. The vital data networks have yet to be developed. Agreement on a usable common language or format, which will allow machines to "talk" to each other, has been slow in coming. MAP has only recently been demonstrated, even though GM began work on it in 1980. Despite all the talk and the "hype," CIM is facing stiff resistance from managers. Companies in the factory automation business – especially GE, Computervision and Cincinnati Milacron – are finding profits hard to come by.

In Britain, two automated manufacturing consultants last year expressed serious concern about progress toward the Factory of the Future. John Leighfield of Istel charged that British executives lacked the expertise and British companies the skilled "systems integrators" needed to put CIM into effect. A survey of 40 companies by Ingersoll Engineers found only patchy progress toward CIM and a poor profit performance among those who had tried sophisticated systems. Brian Small of Ingersoll argued that advanced technological solutions had been "oversold." "Low technology" had in fact a much better track record in boosting profits, and the risks were less onerous, he said. The answer was to "Keep things simple" – like the Japanese.

This conclusion echoes the findings of an earlier (1983) Anglo-German Foundation study of British and West German plants and work at the University of Aston by John Bessant and Keith Dickson, who in 1981 suggested that microelectronics and manufacturing might become "the marriage that never was." This was confirmed by a 1986 British Institute of Management (BIM) survey of 250 UK manufacturing plants, which found a negative pay-off to automation and ambivalent management attitudes.

The second major line of attack on the Factory-of-the-Future scenario is by those who argue that it degrades the quality of worklife by reducing job satisfaction and deskilling jobs. This was a concern of the 1950s writers on automation, and the whole debate gained new impetus with the development of the "human relations" school of management, which stressed the importance of job redesign to improve job satisfaction. Then came the publication in 1974 of Harry Braverman's seminal *Labor and Monopoly Capital: The Degradation of Work in the Twentieth Century* (Monthly Review Press, New York, 1984), which spelt out the Marxist "labor process" argument that the workplace is essentially a battleground between capital and labor. Every effort to introduce new technology and redesign jobs is seen by Braverman as an attempt by employers to increase their control over the workers in order to exploit them further. These general arguments will crop up again in chapter 9.

Very much in the critical, Bravermanian tradition, two US writers have recently directed their fire toward the automated factory. In his book, *Work Transformed: Automation and Labor in the Computer Age* (Holt, Rinehart & Winston, New York, 1985), Harley Shaiken – who was a member of the OTA team looking at factory automation – argues that the new jobs being created in future factories are often every bit as tedious, high-paced and stressfull as old assembly-line jobs. His studies found that machine tool operators had been "deskilled" by the removal of their decision-making power to office-based programmers. Employers were now looking for *less* skilled workers to "mind" machines. Central, computerized monitoring systems keep an eye on work progress, further eroding the skilled machinist's autonomy.

Managers, Shaiken says, are obsessed with the desire to exert tighter control over production by removing skill from the workers and transferring it to complex machinery. This, he says, has hidden costs, because "degrading the work people do ultimately demeans their lives – a cost that is seldom figured in calculations as to which system is more efficient." Nevertheless, "Computer-based automation holds extraordinary promise for improving life on the job."

In a similar vein, David F. Noble, in *Forces of Production* (Alfred Knopf, New York, 1985), argues that attempts to increase productivity by the introduction of new technology are simply part of the ongoing power struggle between management and workers on the shop floor. Modern managers, he says, are particularly keen to destroy the power of skilled machinists.

The trouble is with these one-sided accounts is that they do not stand up to scrutiny. Surveys have shown that most engineers prefer working with CAD systems. In many areas, such as the body-welding sections of automobile plants, the working environment has clearly improved with robotization: who can deny that today's robot "turkey farm" is preferable to the old-style "jungle" where human spot-welders had to crawl over car bodies with weld-guns on long leads? Even Shaiken had to admit, for example, that machine tool operators in his study "generally preferred" CNC machines and that the new technology tends to make work "cleaner and physically lighter."

Two prolific British critics, Professor Howard Rosenbrock of the University of Manchester Institute of Science and Technology (UMIST) and Mike Cooley, author of *Architect or Bee?* (Langley Technical Services, Slough, Berkshire, England, 1980), have similarly had their "deskilling" arguments undermined somewhat by recent studies. For example, Rosenbrock recently told the British Association that the way technology develops "leads to the rejection of human ability and skill." But a study for the Engineering Industry Training Board found that CAD had not reduced the status of the drawing office. A 1986 study by Mark Dodgson of the Technical Change Centre in London found that machine operators *preferred* CNC, saying it gave them greater work satisfaction. Cooley's wild speculations about employers increasingly turning to "mentally handicapped" people have not been borne out by events.

Such critiques tend to see a conspiracy where there is none. Academic engineers have developed new systems in order to improve productivity, not to get one over on the workers. Contributors to two notable British symposiums, *Information Technology in Manufacturing Processes*, edited by Graham Winch (Rossendale, London, 1983) and *Job Redesign*, edited by David Knights, Hugh Willmott and David Collinson (Gower, Aldershot, 1985), steer a middle course between the crude conspiracy theory of some "labor process" arguments and the naive belief that all changes are being made with the interests of the workers at heart. In general, they conclude that employers are not by any means simply motivated by a desire to "control" their workforce.

The choice of new technology is dictated by competitive market pressures and traditional managerial ideologies as much as by power relations in the workplace. Some deskilling may occur, but this need not necessarily be so: "There is no single tendency towards deskilling or reskilling," writes Winch. There is a great deal of room for manoeuver or "strategic choice" as to which type of technology to adopt and how the work is to be organized. It is important to understand the complexities, contradictions and potential of redesigning jobs for the Factory of the Future.

7 The Electronic Office

Now It's the Office of the Future – Convergent Technologies – Dreams and Realities – Slow Progress with Productivity – The Changing Pattern of Office Work – The Automated Office: Prison or Paradise?

Today more people work in offices than in farms, factories, shops and services combined. Over 50 percent of the US labor force (or about 55 million people) are now white-collar workers. In 1900 the figure was only 5 million, or less than one-fifth of the working population: at that time, two and a half times as many people worked on the land as worked in offices. Of the 55 million or so white-collars in the USA today, about 32 million are clerical workers, around 18 million are professional and technical workers and the rest are managers or administrators. For millions of Americans, white-collar work is increasingly the norm.

White-collar workers are engaged in the processes of gathering, storing, manipulating and transmitting information in one form or another. As society grows more complex, the amount of information required to help run it increases. And as the cost of processing this information has come down – thanks to the advances in microelectronics, computing and telecommunications that have been described – the installed base of information-processing technology has increased. Together with the convergence of the computing, telecommunications and office products industries, this tidal wave of new technology is bringing about a major transformation in the place where white-collar workers work: the office.

This fundamental change marks the transition from traditional paperwork to the electronic office, a change that hardly seemed possible ten years ago, when the first personal computers went on sale. Yet installed computing capacity in offices keeps doubling every two to three years, and by 1990 there will be roughly 1000 times as much computing power in offices as there was in 1970. In less than a decade, the US population of keyboards – electronic typewriters, dedicated word processors, desktop personal computers and computer terminals – has grown to nearly 50 million. By 1990 there will be about 110 million electronic terminals in place – or about two for every white-collar worker in the USA.

Now It's the Office of the Future

Apart from technology-push, two forces have been moving things along toward the "Office of the Future." First, there is the pressure on companies to raise office productivity by reducing labor costs, which account for about four-fifths of total office costs. One figure often cited in the 1970s was that, while factory workers had about $25,000 worth of capital investment behind them, office workers were supported by only about $2500 worth of equipment. The potential for automating office work was apparently enormous. Another handy statistic (nobody seems quite sure where it came from) was that industrial productivity had increased by 85 percent in the 1970s, but office productivity had barely crept up by 4 percent overall. Poor office productivity was seen as a major burden on the rest of the economy and a contributory cause of low economic growth.

It's not surprising, therefore, that governments became interested in the productivity potential of office automation. In 1981, for instance, the UK government announced plans to spend a total of £80 million over four years to hasten the electronic office. R&D grants were made available, studies were commissioned, and a series of office automation pilot projects were started. European governments, led by the French, followed suit. In the USA, the Office of Management & Budget has been actively promoting the use of IT *in* government. In 1985 it published new guidelines for agencies like the Securities & Exchange Commission and the Patent & Trademark Office, which want a slice of the federal government's $14 billion annual IT budget. First the "paperless" office – now it's paperless government!

The second factor was the massive marketing campaigns launched by major corporations to sell the Office-of-the-Future concept. Foremost among these were Xerox, Wang, Exxon, ITT, Olivetti, Siemens and Philips, though recently IBM, Apple and AT&T have taken the initiative as personal computers have grown in popularity. At the same time, we have seen a whole variety of computer services, telecoms and office products companies such as the GM-owned EDS converging on the office automation market – the telecoms companies basing their Office-of-the-Future strategies on the PBX, and the computer companies tending to base theirs on distributed data processing or local area networks (LANs).

The drive to mechanize office functions and to reduce office labor costs can in fact be traced back to 1873, the year that US gunmakers E. Remington and Sons introduced the first typewriter. By the late 1890s scores of companies were making typewriters, and by 1900 more than 100,000 had been sold. With the typewriter came an increase in the amount of correspondence and the size of offices. The typewriter also brought women into the office for the first time. Apart from the introduction of the electric typewriter in 1935, basic typewriter technology stayed much the same for decades. Even more recent developments like the "golfball" typewriter of 1961 and the memory typewriter in 1964 were evolutionary rather than wholly revolutionary.

Meanwhile, various attempts were being made to improve office productivity with new devices. Innovations such as ticker tape, automatic telephone switching, calculators, dictation equipment, duplicating machines, offset printing presses and photocopiers made regular appearances throughout the century, but their cumulative impact on productivity was not great. Key items such as that other Victorian invention, the telephone, remained essentially unchanged. Offices were much the same sort of labor-intensive hives of activity that existed at the turn of the century.

The big changes started with the invention of the electronic computer and developments in data processing. The first mainframe computer for business use was installed in the USA by General Electric in 1953. In Britain, the famous LEO (for Lyons Electronic Office) data processing computer was used in the 1950s to process orders from Lyons corner teashops across London. By the 1960s most large companies and organizations had data processing centers for dealing with routine tasks such as the payroll, issuing checks, controlling inventory and sending out bills. The replacement of bulky, troublesome valves with solid-state circuits made mainframe computers cheaper and easier to use.

The advent of the microchip brought forth an avalanche of new products, such as programmable calculators, word processors, facsimile machines, remote terminals, electronic switchboards, sophisticated copiers, microcomputers, minicomputers and super-minicomputers. The humble mechanical calculator, for instance – used for individual and often complex sums rather than the routine, repetitive calculations done by the mainframes – was displaced overnight by microelectronics-based products, which were cheaper, more reliable and much more sophisticated. Word processors – intelligent typewriters consisting of a keyboard and a visual display terminal (VDT) linked up to a microcomputer – were an early success, while "smart" telephones, switchboards and copiers rapidly became commonplace.

Many of the predictions made in the late 1970s about the pace of developments in office automation were wide of the mark. Some of the more improbable scenarios about totally automated paperless offices becoming the norm by 1985 never came to pass. Progress toward the Office of the Future has been slower than expected, and the productivity gains have all too often been small or non-existent.

A British study by OASIS, for example, published in June 1985, found that fewer than half of the UK's 9.2 million office workers had any kind of electronic assistance. The nearest most employees got to automation was the telephone, the calculator and the coffee machine. Even so, expenditure on office automation in Britain quadrupled over the period 1980–4 to £2 billion, and sales of key products like electronic typewriters were growing by 50 percent a year.

The electronic office revolution is also happening in ways that weren't envisaged, the most remarkable development being the rapid penetration of offices by the personal computer. Company bosses, beseiged by equipment suppliers offering a bewildering choice of integrated systems, simply started taking their personal computers into the office instead. Stimulated by the launch of IBM's PC in late 1981, the so-called "desktop revolution" brought 8–10 million personal computers into US offices in just four years. In 1986 the installed base of IBM PCs represented roughly three times the processing power of its 370/30XX mainframes. US shipments of personal computers were about two-thirds the value of mainframe computer shipments.

Thanks in part to the availability of integrated software packages, the personal computer "workstation" or "electronic desk" has become the basic building block of the Office of the Future and suppliers of complete systems like Xerox and Exxon have missed out. The arrival of the personal computer has changed the Office-of-the-Future picture so dramatically that some specialists now talk about "professional personal computing" rather than office automation. The market for professional personal computers is the scene of the fiercest marketing battles, the most savage shakeouts and the greatest success stories.

Managers are also finding that personal computers are remaking their jobs, transforming the way offices actually operate. Putting more computing power and thus information directly in the hands of executives means that companies can dispose of some lower layers of bureaucracy. The onset of professional personal computing can thus lead to major changes in the corporate structure itself and to a general flattening of the managerial pyramid. More and more companies are bringing their data processing, management information systems (MIS) and telecoms units together under one roof – a kind of managerial response to technological convergence.

The transition to the electronic office *is* proceeding, albeit at a slower pace than was earlier envisaged. The installation of information technology in offices (even if not of the type makers originally had in mind) is slowly improving efficiency by reducing the amount of information that is inaccessible at any given moment (the "information float"), by eliminating repetitive tasks such as retyping and filing, and by allowing better, faster information retrieval and decision-making. The electronic office is not without its critics, however: there are those who fear a loss of employment opportunities, a deterioration in workplace relations, a decline in job satisfaction and the creation of new occupational health hazards, chiefly by VDTs. Others point to an increase in nonproductive work such as excessive redrafting and electronic junk mail. But whether the all-electronic office of the future is viewed as a prison or paradise, there are few who wish to see a return to the bad old days of the Office of the Past.

Convergent Technologies

Until recently, most of the changes in office equipment had been gradual enhancements of existing products: superior typewriters, copiers that could produce better images, improved dictation machines and so on. This long period of *mechanization* involved no fundamental change in traditional office technology – innovators were simply mechanizing manual tasks. The office "system" was still based on shuffling paper – usually 8½ in × 11 in. pieces of paper, which fitted neatly into typewriters, filing cabinets and facsimile machines.

All this changed when the advent of microelectronics in the early 1970s got designers thinking in terms of office *automation*, which could be made possible by the growing digitization of information. The digital revolution means that voice, data, facsimile and video can be reduced to a series of electronic pulses, while the notion of "automation" encourages people to look at things like offices as so many "functions" (for example, the "keying function," the "filing function," the "mailing function," etc.) of a total system. Digitization would thus be the glue that would hold together all the elements of the modern "office information system." Strings of numbers would replace paper as the system's working unit.

The migration to digital equipment has been spectacular and swift: in 1970 less than 200,000 remote terminals in US offices could access computers. By 1980 there were over 2 million and in 1984, 12 million. The total number of digital devices topped 20 million in early 1984 and should reach over 100 million in 1989. Digital is overtaking analog: the total of analog devices, primarily telephone handsets, stood at 52 million in 1984, but will grow only to about 64 million in 1989.

Technological convergence and digitization have thus brought computer, telecommunications and traditional office product companies converging on the electronic office market. Computer companies like IBM now find they also have to be in telecommunications (although "Big Blue" kept its office products and computer divisions separate for a long time). Other computer companies, such as Honeywell, NCR, DEC and Sperry, were late into microcomputers and/or word processing and to some extent missed the boat. US telecoms companies like AT&T have tried to get into computers, while foreign-based companies like Sweden's Ericsson, West Germany's Siemens and Japan's NEC found they had to offer a total range of computer and office products, from complete systems right down to tiny calculators and office furniture.

Five or six years ago, office automation was virtually synonymous with word processing. But ever since IBM introduced its PC, office automation has come to mean end-user computing. However, word processors are still a major office product – one of the "big four" markets, along with personal computers/workstations, copiers and PBXs. Over 4 million word processors had been sold in the USA by 1986.

A word processor is an intelligent typewriter, consisting of a keyboard and VDT linked to a computer. Each keystroke operates an electronic switch rather than a mechanical linkage, which means that information coming from the keyboard can be transmitted in digital form to an electronic memory. The great advantage of a word processor is that it abolishes retyping: material can be manipulated, edited and formatted on the screen before being printed out as hard copy or committed to memory. Word processors have proved especially useful in companies and organizations whose work involves standard letters, regular mailings or the production of legal documents, builders' specifications or operating manuals.

Once it seemed that word processors would completely oust electronic typewriters. But in fact, these humbler machines have undergone a dramatic renaissance, with the help of some nifty marketing by Japanese and European suppliers. The installed base of electronic typewriters in the USA reached about 1.5 million in 1986, compared with about 4 million stand-alone and clustered word processors. Lower prices mean that US sales are growing at a faster rate (about 33 percent a year) than word processors (about 25 percent). In the UK electronic typewriters are outselling word processors in absolute terms by an even greater margin.

In an effort to beat off the challenge from personal computers and multifunction workstations, word processing vendors are adding personal computer functions, color graphics and enhanced peripherals like laser printers to their products. This is resulting in a further blurring of the distinction between "typewriters" and "computer terminals" which is symptomatic of the overall convergence process.

Image processing in the office is dominated by the stalwart office copier. By 1988 there will be over 4 million of these installed in US offices, although the rate of increase in sales has recently slowed. Suppliers like Japan's Canon, Ricoh and Toshiba are making major inroads into Xerox's traditional territory, especially at the cheaper end. Copier technology is improving all the time, making possible higher-resolution images, reduced costs and the production of smaller desktop machines. Some models have even been designed for home use.

The "smart" or intelligent copier is microprocessor-controlled and is capable of automatic exposure-setting, enlargement and reduction, the visual display of operational information, and monitoring of use. And stand by for the laser copier: unlike the conventional copier, this one scans the original image and breaks it down into digitized, electronic signals which modulate a laser beam in the printing unit. But once the image has been converted into electronic form, it can be manipulated, stored and transmitted like any other digitized information. The next generation of intelligent copiers will thus be able to communicate with word processors, computers, databases and remote terminals – perhaps miles away – to provide a fully integrated information-processing and printing system.

The dramatic arrival of personal computers and executive workstations in the office also illustrates this process of integration and convergence. Personal computers can be used for word processing, number-crunching and spreadsheet analysis. They can be hooked up to mainframes, fax machines, printers and databases, transforming themselves into what have been variously described as "multifunction workstations," "executive workstations" and "the electronic desk."

Spreadsheet analysis, made possible by sophisticated new software packages, is growing in popularity. Beginning with VisiCalc in 1979 and continuing through Lotus's 1–2–3, Symphony and Jazz, spreadsheet packages have given professionals a whole new way of working. Executives are able constantly to monitor company performance: at their fingertips (in theory) are all the information and variables they need to plot future business strategy and experiment with different scenarios. More than a million spreadsheet packages were sold in the USA in 1984 and their fans claim that they provide a sophisticated and flexible weapon for tackling the competition. Critics say they are not so useful and are at the mercy of that old truism, "garbage in, garbage out."

Until recently, most workstations were used for a single, dedicated purpose, such as word processing or accessing databases. But increasingly, the uses of workstations have been extended by new devices and software to the point where workstations have become central to the office information system and to the working lives of executives and professionals. Through this communications "hub," the busy professional can keep in touch with other people and vital sources of information. Most new integrated office networks are being built around workstations.

Various factors are holding things up: the slow speed with which good software is being developed for linking micros to mainframes and databases; problems with software for specific tasks such as accountancy and storing legal documents; failure to agree on software operating systems; and delays in the development of the crucial local area networks (LANs) to link up all the personal computers/workstations in an office (or factory, hospital or university), thus enabling employees to share expensive, centralized resources. LANs are now generally considered to be the foundation upon which the Office of the Future is being built, and the Japanese in particular are going crazy over them.

But progress with LANs has been bedevilled by high costs and the failure to agree on standards. There are three major types of system (the "bus," the "ring" and the "star") and three types of cable in use (coaxial, twisted pair and optical fiber) – and vendors continue to wrangle about the merits of each. Xerox's Ethernet, first demonstrated in 1973 and marketed by DEC, is still the best-known. It faces competition from Wang's Wangnet and especially from IBM, which finally entered the market in October 1985 with its own "token passing ring" system developed in conjunction with Texas Instruments.

Because of the delays and uncertainty, some companies have turned to telecoms companies like Rolm and Northern Telecom, offering modified PBXs as the basis for simpler, cheaper LANs. But with IBM's entry, sales of LANs are now rising rapidly and the market is expected to be worth $2 billion a year by 1988.

While workstations have advanced to center-stage, computer technology has responded to the challenge. For example, a new generation of more powerful microcomputers called "supermicros" are taking the place of old-style minicomputers. Supermicros incorporate standardized high-performance microprocessors which vastly outperform ordinary micros and cost much less than minis. Unix is rapidly becoming the standard operating system for these machines – as indeed it is for many office products.

Better storage systems based on floppy disks are being developed all the time. But most of the excitement recently has been generated by optical processing and optical disks, which threaten to replace magnetic disks. Optical processing technology uses lasers to read the digital code embedded in optical storage disks. These disks can store massive amounts of data at low cost. Digital optical storage disks are a spin-off from work on videodiscs and the compact audio disc (CD) – hence their name, CD ROMs.

While computers and office products have come to resemble one another, telecommunications products are converging from another direction. As digital technology takes hold, office telecommunications are being transformed more rapidly than at any time in history – a process stimulated by the impact of deregulation. "Smart" telephone handsets have become commonplace, while the market for digital PBXs has exploded.

PBX's have gone through four generations in less than 15 years: the latest models can serve as the voice and data hub of the entire office network. Upstart companies like Mitel and CXC have rushed out new products which embrace data switching as well as sophisticated call management. Call management can help companies contain telephone costs by, for instance, providing a printout of all calls made; call blocking, forwarding and prioritizing; and actively seeking the lowest cost route for calls.

Meanwhile, the old slow and smelly telecopiers have given way to high-speed digital fax machines. By the end of 1985, Japan had 850,000 fax installations, the USA 550,000 and Europe, 120,000. Cheaper and better-quality devices should rapidly boost demand. Electronic mail in all its forms, audio and video teleconferencing, videotex and satellite communications – as we saw in chapter 4 – are all part of this business technology explosion.

Nevertheless, the digitization of telecommunications will not be complete for some time. By 1990 about 90 percent of metropolitan services and toll switching will be digitized, but only about 30 percent of intercity services and local switching will be fully digital, and the local loop – or "last mile," as it has been called – is proving such a bottleneck that companies are indulging in various strategies involving microwave, cable and DBS in order to "bypass" it. Although it may take many decades to replace all the old copper wire in the telephone network, by 1990 the basic infrastructure should be in place for ISDN (the Integrated Services Digital Network). This will give a further boost to the electronic office.

The boundaries among office products and office technologies have thus become blurred as the various components of the Office of the Future have become interconnected. This will become more so as new devices become available, such as optical character readers (OCRs), which can scan pieces of paper and convert the information into digital code. Communications will be vastly improved if "speech recognition" machines, or so-called "talkwriters" that convert speech into text, ever go commercial. It's too early to tell for sure, but experiments at MIT and Carnegie-Mellon, plus prototypes developed at IBM and by start-up companies like Kurzweil and Speech Systems, suggest that voice processors which eliminate the need to key-in type may one day be obtainable. Now that *would* be a revolution!

Dreams and Realities

> "Tell me, then, what you think of the . . . " There Dr Breeze paused, took a long breath, and puffed himself up to say in a stentorian voice, "the Office of the Future"?

Perhaps it was the gathering blue dusk, the sense of disorientation created by the cluttered room, or the jet lag, the tiredness, the mental strain. Anyway, I heard myself say:

"There isn't going to be an Office of the Future."

"Oh!" said Dr Breeze in mock astonishment. "You must not let anyone hear such words." Blushing, he went to the door and closed it carefully, after looking to the left and to the right to make sure no one had heard my statement. When he came back, he spoke low, in conspiratorial tones.

"Who are you, some kind of anarchist? Everybody is working on the Office of the Future, everybody is investing in it, and now you're saying it isn't going to exist. They'll kill you if they find out."

Imaginary conversation between Jacques Vallee and Dr Breeze, the Wise Gnome of Washington, in Vallee's book, *The Network Revolution: Confessions of a Computer Scientist* (Penguin, Harmondsworth, England, 1984)

In 1975 oil giant Exxon, keen to diversify out of the oil business, plunged into the office automation market with the purchase of start-ups Vydec (word processors), Qwip (fax machines) and Qyx (electronic typewriters). This was Exxon's bid to become a major force in what was seen to be the growth industry of the decade. During the next ten years, Exxon poured $2 billion into its Exxon Office Systems subsidiary, spending lavishly on advertising and promotion. But it never made money, and early in 1985 Exxon sold the assets of the business to Lainer Business Products for a trivial sum.

Exxon's drive to disaster in the office was mirrored elsewhere: the UK's National Enterprise Board in the late 1970s poured millions of pounds into Nexos, a company that was to be the British entry into the Office-of-the-Future stakes, but it collapsed with huge debts in 1981. Logica, the UK software group which was linked to Nexos, pulled out of the office automation market at the end of 1985.

The story of office automation's slow progress goes back to 1971, the year a team of scientists at Xerox's research center in Palo Alto, California, began a project to design what they called "the electronic office of the future." The idea of using digitization to create integrated office information systems was sound enough, but what was missing at the time was the technology to make the dream a reality (the human and organizational problems were to emerge later). But the idea was a tremendously powerful one, and soon companies, governments and academic research establishments all over the world rushed to develop their own Office-of-the-Future scenarios. There was a great fear that companies or countries not in the office automation market would miss out on a potentially huge market.

Xerox stayed ahead for most of the 1970s, closely followed by Wang Laboratories of Boston, the company founded by Chinese-born Dr An Wang. Wang was particularly strong in word processors and small business computers. By 1980 the rush to get into the electronic office market became a stampede, with established international companies like the Dutch group Philips, France's CIT Alcatel and Japanese suppliers like Matsushita well to the fore, as well as (apart from Exxon) complete outsiders like West German car maker Volkswagen, which purchased Triumph–Adler, the typewriter company.

Everyone awaited the arrival of IBM, which someone likened to a killer whale circling the office automation market waiting to spot the rich pickings. In 1981 IBM struck, with a string of new products including its PC, which was to spearhead the attack. For several years, from about 1977 to 1981, Apple and Tandy/Radio Shack had pretty much had the business personal computer market to themselves. But by mid-1982 about 120 personal computer vendors had entered the fray – some becoming overnight millionaires – and IBM's PC was selling by the trainload. Bulk purchases of PCs by corporations boosted IBM's market share to nearly 30 percent by 1983. Apple, which meanwhile had targeted the office/business sector for future growth, went "head to head" with IBM, although its 14 percent market share in 1983 was half that of IBM.

At the same time, a lively battle developed in the fast-growing market for business software packages. By 1983 IBM had not achieved the same dominance in this sphere as it had in hardware, having only a 9 percent market share, the same as Tandy. Close behind with about 3–5 percent were no less than nine other software producers: Apple, Commodore, MicroPro, Microsoft, Lotus, VisiCorp, Digital Research, Ashton-Tate and MSA (Peachtree). Literally hundreds of others shared the remaining 46 percent of the market.

As IBM began to dominate the electronic office, "IBM-compatibility" became a must for all vendors except Apple. Many other computer manufacturers standardized on IBM's MS-DOS operating system, which was a necessary, but not a sufficient, condition for survival. IBM also undoubtedly benefited in the process by which MIS managers once again took charge of company computer-purchasing decisions. With the amount of corporate cash being spent on personal computers and their add-ons rising rapidly, companies became more conservative and thus more likely to opt for IBM products. This particular sector of the office automation market had followed the classic process of initial gold rush, major shakeout and eventual consolidation around two or three dominant suppliers. As Frost & Sullivan pointed out, it was also clear that the days of the separate "personal computer" market were also numbered, as it merged to become part and parcel of the office systems market.

Recently, the overall office automation market has begun to look increasingly like a one-horse race involving IBM. But the transformed Italian office equipment company, Olivetti, has also made a major play for the office automation market, tying up with AT&T and Japan's Toshiba, and launching a new range of products such as its video-typewriters. Apple in 1985 tried once again to turn the "Blue" tide with a massive campaign to get the Macintosh and the Lisa (renamed the Macintosh XL) adopted as standard business machines – but found it tough going. Despite Apple's continued efforts, IBM kept gaining market share in personal computers. Meanwhile, Compaq, Northern Telecom and other manufacturers were tempting business customers with a new breed of "computerphones" – machines that combine telephone terminals with personal computers.

However, despite all the marketing hype and the success of certain products like the business personal computer, it was apparent by the mid-1980s that the Office of the Future had not yet arrived. The expensive, integrated systems had failed to sell, and most personal computers in offices were not linked up. Surveys in the USA and Europe showed that office automation still had a long way to go. Patricia Seybold, editor of the widely quoted *Seybold Report on Office Systems*, said that the promised land was still ten years away. A British authority put the state-of-the-art at "about two and a half out of ten, if ten is what is being marketed."

To be fair, doubts about the Office-of-the-Future scenario had crept in early on. Skeptics pointed out at the time of the great office automation rush in 1980 that this particular market was peculiar in that it was to be manufacturer-driven, not user-driven. By 1982 it was obvious that the expected sales bonanza had not materialized, although spending was rising. Articles began to appear analyzing the reasons why the office revolution had been delayed.

Moreover, there was growing evidence that some executives and professionals were getting disillusioned with computers. Too many had been sold the "wrong package" of software, and many had suffered bad experiences with faulty and/or incompatible hardware. PBX vendors found that customers were not actually using the "smart" features on their devices. The buying spree of the early 1980s had resulted in a lot of expensive gear lying around gathering dust: surveys showed that even personal desktop computers were badly under-used and in some instances had been completely discarded. This has produced outright disenchantment among some, and a more cautious appraisal among others that the Office of the Future won't have arrived by tomorrow.

Slow Progress with Productivity

The slow progress toward the Office of the Future has been caused by a variety of structural, financial, technological and human factors. One very basic – but often overlooked – reason is that there has simply been a lack of suitable buildings to receive the new technology. Existing office buildings were not designed to cope with the masses of wiring required and the amount of heat generated by the new automated equipment. Traditional skyscrapers are not flexible enough for the new working environments. In the UK, the Orbit report in 1983 warned that the need to rethink office design was urgent and that the cost of remodelling old buildings to take the new technology could be almost as high as to build new.

The answer seems to lie in the construction of more "smart" office buildings which are better suited to the infotech age. These not only contain ample ducting for communications networks, but can also be made "intelligent" with the incorporation of all the latest energy control and security monitoring systems. Examples already exist in Dallas (Lincoln Plaza) and Los Angeles (Grand Financial Plaza), as well as in Chicago, Denver, New York City, Arlington (Virginia) and Hartford (Connecticut). But in the short term, many companies will have to make do with false floors (like the Toshiba system) and flat-cabling (like Versa-Track), without which the installation of vital networks would be even more difficult than it is already.

Aside from the colossal expense of most Office-of-the-Future products – especially in the early days – progress has been further slowed by manufacturers pursuing technological cul-de-sacs, shortages of suitable software and the lack of agreement on standards. In such a fast-moving market, vendors often try to gain advantage by designing unique products. But uniqueness leads to incompatibility, and incompatibility prevents the explosion in demand, so the market fails to achieve its full potential. This contradiction means that both customers and competitors can suffer. The classic example is in LANs, where frequent attempts in the USA and Europe to get manufacturers and other organizations to agree on standards have failed and we shall now have to see how the issue is resolved in the marketplace. Electronic mail has been similarly stalled.

The inability of machines to communicate has proved to be a major stumbling block and has put off many potential customers. Too many companies have reason to regret buying different brands of computers that can't "talk" to each other. And because manufacturers have got into the habit of announcing new products far in advance of their actual introduction, customers have learned to postpone purchases until the new models arrive – or, in many instances, until prices come down. This has further slowed the pace of change. Confused by so much choice, customers suffering from "option shock" have also wanted more time to digest what they have already got before plunging into new purchases.

Management indifference has also played its part in delaying the electronic office. Too many managers prefer to carry on in the same old way and won't explore new methods of working. They feel threatened by the new technology and fear loss of power to those more familiar with computers. "Wait and see" conservatism all too often triumphs. In the UK, a major survey by the Policy Studies Institute in 1983 on progress and problems with the electronic office found the biggest single reason for not investing in new technology was that it had simply never been considered. Many managers lacked the knowledge or commitment to even investigate the possibility of going over to automation. Management resistance in the 231 offices surveyed had been far more prevalent and more effective than any resistance on the part of the office workforce.

The fear of typing on the part of executives was noted long ago. But in the early 1980s a new condition – "terminal phobia" – was diagnosed to explain the unwillingness of executives to use keyboards. This was caused partly by a feeling that their use was demeaning and partly by the fear that they would be shown up to be ignorant and unfamiliar with new technology in front of their staff.

The *psychology* of office automation has thus come to the fore, as it has been increasingly realized that human factors are just as important as the technology. Managers are unwilling to change, and feel threatened and confused. Those lower down the office hierarchy see the potential disruption of traditional working methods and perhaps feel, too, that their job is on the line. Power battles have developed between different company departments; in particular, the technical experts in the data processing and management information services (MIS) departments have struggled to reassert control over purchasing decisions, after managers and professionals started taking their own computing equipment into the office. MIS departments have had to respond to end-user pressure for personal computers, better back-up and improved networking.

Some argued that the problem was a wider one of the technology outstripping the ways of organizing it. Developments in office technology had raced ahead of the human ability to comprehend and manage the complexities. What was now needed was a period of learning and consolidation, so that users might learn how best to use the equipment they had already installed.

With this in mind, the UK government in 1981 launched 20 office automation "pilot projects," which took the form of two-year trial installations of new systems in a variety of public and private organizations, including the UK government itself and the BBC. Some were more successful than others. A series of bulletins on progress were published through 1984 and 1985, as part of the overall learning process. A joint conference was held, survey reports were published, and a campaign was launched to spread the lessons of the pilot projects. The main lesson learned was that office automation is not a panacea and will not revolutionize offices overnight. Automating office work is far more complex than anyone had imagined, and future progress will be slow – but there *will be* progress.

Probably the most important factor slowing progress toward the Office of the Future is the continuing fact that productivity gains are not easily demonstrable. White-collar "productivity" is not amenable to precise measurement like manufacturing productivity, and therefore assessing the possible gains has been extremely difficult. The claims made by vendors have often proved illusory, with the result that improved office productivity has become a kind of elusive Holy Grail.

Measuring office productivity ought to be easy. Surveys of how office workers spend their time show clearly that long periods are spent on the phone (often playing "telephone tag," that is, trying fruitlessly to locate people), in seemingly endless meetings, simply wandering around or apparently doing nothing. Since white-collar salaries cost the US over $1 trillion a year, even a tiny percentage gain in productivity would yield billion-dollar savings – in theory.

In theory, the potential for raising office productivity is enormous. In practice, things ain't so straightforward. The substitution of machines for human labor does not increase office productivity as straightforwardly as it does in a factory. There is no clear link between spending money and seeing results. Even measuring the output of word processors is difficult, and the studies that have been done have yielded inconclusive results. A survey of 4000 offices by SRI International found that, in all but a few special cases, such as legal departments, there were few direct cost savings from any kind of automation, including word processing. A summary of British word processing research by the Manpower Services Commission concluded that most systems were not being fully utilized because operatives lacked effective training. Centralized "word processing centers" have mostly been abandoned as failures. Patricia Seybold says: "The organizational benefits of word processing are so intangible that you can't get hard proof of dramatic savings."

Where productivity gains are discernable, for example as was seen in the well-known study of Hanscom Air Force Base, the time saved by office automation is not always used wisely. Sometimes the new technology may well boost output, but simply increasing the amount of paper or data in circulation (for instance) may be of dubious benefit. In fact, many executives are already complaining that they suffer from "information overload" and "info-glut."

When it comes to measuring managerial and professional productivity, the task is even more tricky. If executive productivity could be boosted, few doubt that the rewards would be substantial: according to one estimate, professional and managerial salaries actually account for 68 percent of total office salary costs. The problem is that much of an executive's time is spent on interpersonal communication, policy development and dealing with problems that are not amenable to automation. However, the quality of executive decision-making affects the performance and the future of the entire enterprise. Any improvement here brought about by easier access via computer to the relevant information will have an enormous –if immeasurable – impact on the company's fortunes.

The Changing Pattern of Office Work

Concern about the impact of new technology on office work has focused on four main issues: the likely decrease in employment opportunities; the quality of the jobs remaining; possible health and safety hazards associated with VDTs; and fears that computerization will have an adverse effect on social relations within the office. As is often the case, observers in these matters generally divide into optimists, who see the wonderful world of office automation through rose-tinted screens, and the pessimists, who gloomily argue that automation will degrade the quality of office worklife, reduce job satisfaction and deskill jobs – in other words, they suggest that the Office of the Future will be no better than the factory of the past. The truth, of course, lies somewhere between the two.

Fears that office automation would mean the sudden and widespread loss of jobs have not been borne out by events. In the late 1970s and early 1980s, a number of reports appeared in the USA predicting massive white-collar job losses. In a famous article, Bell Laboratories executive director Victor A. Vyssotsky pointed out that the potential for labor-shedding in offices was enormous: a 2 percent reduction in office staffs would displace 25 million workers by the year 2000. In Britain and Europe, the "mighty micro" hysteria produced some wild and completely unsubstantiated predictions that offices would "soon" become deserted as the services of typists, filing clerks and secretaries would "soon" no longer be required.

Even the more sober analysts focused on the loss-of-jobs angle. In Britain, for example, the Equal Opportunities Commission published a report in 1980 predicting that "up to 170,000" UK typing and secretarial jobs would be lost by 1990. Incomes Data Services reported in 1981 that the arrival of word processors and the like had *already* led to the net loss of employment opportunities in the companies studied. In 1982 a Science Policy Research Unit study of women's employment prospects reported that computerization generally led to a loss of routine clerical jobs and the creation of a smaller number of more skilled, technical jobs. But this phenomenon was being obscured by the growth in office work as a whole.

Perhaps the most wide-ranging survey of what has been happening on the employment front so far is Diane Werneke's *Microelectronics and Office Jobs*, published by the International Labour Office (Geneva) in 1983. Werneke found that new office technology was labor-saving, but had not yet resulted in widespread labor displacement in the OECD countries studied. Rather, it had been reflected in a slowing of demand for certain categories of workers, especially routine clerical workers – a point reinforced by a UK Institute of Manpower Studies report published in 1985. The employment impact of new technology had also been muted by the slow rate of diffusion to date. And the rising demand for information and information services had further obscured the reduction in the labor content of certain tasks (Forester (ed.), *Information Technology*, 1985).

Quite simply, the introduction of new office technology generally allowed more work to be done by the same number of or fewer people. Particularly in the banking and insurance fields, reduced recruitment, natural wastage and unfilled vacancies were the methods used to adjust manpower levels. As to the future, the pace of change was hard to predict as there were two conflicting views of how economic growth might affect the demand for new technology: one is that low growth increases competitive pressures and thus the desire to boost productivity; the other is that low growth dampens down investment in new technology.

While the net employment effect of new office technology is to reduce the labor content of certain office tasks, most studies emphasize the restructuring of office work: the proportion of jobs requiring technical training and higher skills is increased, while the demand for clerks, typists and other workers in routine or semi-skilled office jobs sharply falls. A further twist is the enhancement of secretarial skills and the personal secretary's role at the expense of routine copy typists. This is actually a boost to the quality of some women's employment, whereas the loss of routine office jobs has reduced the quantity.

For example, a 1983 study for the UK's National Economic Development Office on the impact of advanced office systems found that new technology either reduced employment or inhibited further growth in most of the companies studied. Jobs lost were the boring ones, such as those of clerks, typists and machine or data processing operators. The jobs gained were more technical and demanded a higher level of discretionary skills and training. They also provided greater job satisfaction, and the extra responsibility provided employees with a broader view of the company's entire operations.

The big fear is that there will not be enough of these higher-quality, skilled jobs to go round. As Brookings and MIT researchers have pointed out, there is little evidence so far that the new high-tech jobs being created will be anything like numerous enough to replace the unskilled jobs being eliminated through automation. In offices, also, we may see a growing polarization between a high-tech elite of executives and technical specialists and a diminishing band of old-style, semi-skilled office workers. Werneke, in her study for the ILO, found that many clerical jobs had been deskilled through the introduction of new technology, with the result that the clerical workforce (consisting mainly of women) was effectively cut off from traditional paths to promotion.

Much depends on the organization of work, the pace of work and the methods of employee control adopted. New technology can be used to decentralize work, giving employees greater autonomy, discretion and responsibility; or it can be used to deskill jobs, to fragment tasks and to increase monitoring and control over employees, which will undoubtedly lead to greater employee alienation of the kind normally associated with factories. The choice lies with the employer: jobs can be computer-aided or computer-degraded. New technology can be used to redesign jobs in such a way as to increase or decrease the quality of working life. Just as there is no uniformity of impact of new office technology, there is no independent and inevitable impact of technology per se.

The Automated Office: Prison or Paradise?

Ever since the use of VDTs became widespread in the 1970s, there has been a steady stream of complaints that they cause headaches, backache, eyestrain and excessive fatigue. A more serious allegation is that the radiation emitted from VDT screens can cause cataracts, miscarriages and birth defects. As the number of employees using VDTs swelled to millions from thousands, labor unions and groups like 9 to 5 (the National Association of Working Women) in the USA began publishing reports of alleged health hazards. In the UK and Europe, a number of leading white-collar unions such as Britain's APEX issued their own guidelines for the operation of VDTs and began to press companies to sign "technology agreements" regulating their use. Some employers suggested that the real motivation behind the concern of the unions was a desire to recruit in the relatively under-unionized office sector.

The first major report on the issue was published by the US National Institute for Occupational Health and Safety (NIOSH) in August 1980. NIOSH expressed "cause for concern" about possible visual, muscular and psychological hazards for those using VDTs over long periods of time. At the three major sites studied, VDT operators reported more symptoms than other office workers, including blurred vision, eyestrain and stiff necks. They also seemed to suffer more from general health complaints and psychological problems such as stress, irritability, depression and anxiety. But NIOSH did point out that high stress levels were not caused solely by the use of VDTs: a lot depended on differences in working environments – specifically, whether employees faced high work demands, tight deadlines and/or a lack of control over their jobs.

In 1981 a National Academy of Sciences (NAS) conference on the subject in Washington reflected the growing consensus that ergonomics (the science of adapting machines to people) can resolve only some of the undoubted problems suffered by regular VDT users. Improved seating, better lighting and ventilation, and especially more appropriate screen technology can do much to reduce backaches, headaches and eyestrain. But there was growing evidence that, overall, job stress was as much a cause of the complaints as the VDTs themselves.

In the UK, Health and Safety Executive reports in 1983 and 1986 specifically cleared VDTs of *radiation* hazards, although a Canadian Center for Occupational Health study was less categoric and even suggested that pregnant women should avoid using them until more was known. As the number of VDTs in use in the USA passed the 10 million mark, the second major US study of the subject was published in July

1983 by the National Research Council. Its 273-page report seemed to clear VDTs of responsibility not only for cataracts and birth defects, but also for less serious symptoms like eyestrain: these, the report maintained, could be cured by ergonomics.

"Our general conclusion," said study group chairman, Professor Edward Rinalducci, "is that eye discomfort, blurred vision and other visual disturbances, muscular aches and stress reported among VDT workers are probably not due to anything inherent in VDT technology."

The NCR study brought forth a critical response from labor unions and groups like 9 to 5. Pressure continued to grow in some states for new laws regulating the use of VDTs, similar to those passed in Sweden and West Germany. Moreover, NRC's findings did nothing to quell further speculation that VDTs were a problem.

In 1985, for instance, a spate of new reports appeared: a study of 13,000 workers by Japan's General Council of Trade Unions found a high level of miscarriages, premature births and stillbirths among women VDT operatives; a New York University professor diagnosed a new stress-related condition, "video blues"; while in the UK a Labour Research Department report found an unusually high level of complaints among VDT users. Even more disturbing, an unpublished report for IBM by Professor Arthur Guy of the University of Washington in Seattle concluded that radiation hazards could not be entirely discounted, and actually recommended an extra layer of shielding on VDTs to prevent possible exposure.

Just to confuse matters, the safety-conscious Swedes, in the form of their National Board of Occupational Safety and Health, published a major study of 10,000 pregnancies which found no statistically significant differences between the pregnancies of women who experienced low, medium and high levels of exposure to VDTs. "We see no medical reason for recommending that pregnant women be exempted from working with VDTs," concluded the Swedish reseachers. But they emphasized that stress at work was a serious problem and a major contributor to ill health.

The fourth area of concern in the office automation debate is that of the possible impact of computerization on the general atmosphere and social relations within the office. This issue is enmeshed with the arguments about deskilling, job satisfaction and job redesign already reviewed. It also embraces various conceptions – and perhaps myths – about what offices were like in the past.

In its most extreme, pessimistic, Bravermanian form, Marxist critics argue that we are about to witness the end of the relaxed informal "social office." In its place will come regimented information-processing areas, rather like modern factories. Capitalists, in their drive to extort more surplus value out of office workers, will seek ever tighter control over the labor process. As Braverman put it, the new office technology aims at "squeezing out the minutes and hours of labor power lost in the personal relations and contacts among secretaries and between secretaries and their principals."

Obviously, there is some truth in this, in that secretaries have been able to carve out an area of freedom for themselves under existing office relations. And according to Alan Delgado's social history of the office, *The Enormous File* (London, 1979), the historical trend in office social relations seems to have been away from the formal to the informal. One hundred years ago, he says, men and women in offices were segregated and talking was often not allowed. The introduction of office automation might herald a return to those "bad old days" of highly regulated work, although now in a machine-dominated environment.

But this argument tends to put a "halo" effect around contemporary paper-shuffling offices, and it presupposes that this so-called "social office" is in fact sociable and popular with office workers anyway. According to a 1980 Opinion Research Corporation survey, US office workers are very discontented, and they particularly dislike the more sociable "open-plan" type of office. And, quite contrary to many of the one-sided critiques and pessimistic union pamphlets on the subject, new office technology and the working environment that goes with it has proved popular with at least some groups of workers. A 1984 UK survey by Manpower Ltd, for instance, revealed that three-quarters of qualified private secretaries thought that new technology would relieve them of repetitive tasks, enabling them to give more attention to interesting, discretionary matters. Far from leading to a decline of the "social office," new technology might give it a new lease of life.

An equally optimistic view is provided by Vincent E. Giuliano, an acknowledged authority on office systems. Giuliano distinguishes three stages in the evolution of offices: the pre-industrial, the industrial and the information age. The first two correspond to the well-known artisan and industrial models of production. Many small businesses are still at the pre-industrial stage, Giuliano says. They have not yet gone over to production-line working of the kind that characterizes the industrial office. The third stage, the information-age office, uses new technology to preserve the best aspects of the earlier stages and to avoid their failings. (Forester (ed.), *Information Technology*, 1985).

In an information-age office, says Giuliano, the emphasis is on people-centered rather than machine-centered work. From the individual workstation, the information-age office worker can generate and retrieve information at his or her own pace. There is no such thing as "work in process" with all its attendant uncertainties. Productivity is improved and so is employee job satisfaction. Instant access to information and a happier staff means greater customer satisfaction. And having more satisfied customers gives a company a major competitive advantage.

Whether the fans or the critics of the Office of the Future are proved right in the long run remains to be seen. The truth is that there is no inevitability about either the rosy or the gloomy scenarios. The technology itself doesn't determine anything. In a study of the introduction of word processing (*Journal of Occupational Psychology*, No. 55, 1982), British authors David A. Buchanan and David Boddy found that changes in the overall pattern of jobs and skills were determined weakly and indirectly by the technology and strongly by management ideas about how work should be organized and controlled.

A major Rand Corporation study concluded that a wide range of work situations is possible using new technology:

> The way the technology is used – and its impact on the lives of the workers – depends as much on management ideology as on the technology itself. The beliefs that managers hold about human nature will guide their decisions on automated systems, and the consequent impact of the technology on individuals and jobs will, in turn, tend to confirm their beliefs (*The Futurist*, June 1982).

In order to create an automated office that is productive as well as a satisfying place in which to work, the Rand researchers recommend extensive user participation in planning for, and decision-making about, the installation of new technology; more and better training in its use; greater anticipation of and preparedness for future changes; and the creation of an adaptive, flexible work environment.

In other words, the future of the office is negotiable. Its final form lies in the hands of today's management and the office workforce, each of whom has just as much right as the others to debate and shape the kind of offices they wish to work in.

8 Micros and Money

Keeping the Customer Satisfied – Electronic Banking: Fast Bucks – Financial Information Goes On-Line – Paying with Plastic – The Retail Revolution – Home Banking and Shopping

When someone writes a check, no money changes hands. What happens is that the information on a check simply causes the banks to alter their customer accounts. Even a banknote only contains the information, "I promise to pay the bearer:" on its own, it is worthless. Money is thus really *information* about money. Indeed, the whole of commerce is based on information – information about goods and services, information about their prices, information about storage and transportation, and information about payment. It's not surprising, therefore, that banking, retailing and distribution are being deeply affected by the information technology revolution.

The cost of processing all the information needed to keep the wheels of commerce rolling is phenomenal. Take checks, for instance: every year in the USA, about 40 billion checks are written and the cost of processing them runs out at about $30 billion. Handling cash is even more expensive for the banks and the federal government. In Britain, the London clearing banks maintain at least £1 billion in notes and coin at any time, while Barclays Bank spends about £15 million a year simply moving cash around between branches and regional cash centers. In a determined effort to boost productivity and cut costs, therefore, US banks will be spending $11 billion a year on computers and automation by 1989 – up from $4 billion in 1984 – and total spending by US banks will amount to more than $50 billion over the next five years.

Banks, in fact, were among the first commercial users of mainframe computers back in the 1950s. This was because routine financial transactions were particularly amenable to automation. The whole notion of "electronic funds transfer" (EFT) predates the chip by almost two decades. But the mid-1960s predictions about the imminent EFT revolution and the "cashless society" being "just around the corner" never materialized.

At that time, the argument was that the banking system was sinking in a sea of paper as the number of checks written rose remorselessly every year. People would "soon" be paying for everything with plastic cards, and EFT would rocket funds around the nation in seconds, thus reducing the growing paper mountain. However, the high cost of equipment like automated teller machines (ATMs), plus resistance to payment-with-cards by consumers (who feared errors, fraud and loss of privacy), a lack of cooperation between banks and a mass of legal problems concerning what bank branches could and could not do meant that progress was slow.

In the late 1970s and early 1980s, however, things started to happen. Thanks to the microchip, the costs of equipment such as cash dispensers and ATMs came down, with the result that over 50,000 ATMs had been installed in the USA by the end of 1985. By then, Japan had more than 40,000 ATMs and France, 10,000. Consumer resistance to EFT waned. While the chip dramatically increased the power of computers behind the scenes, with the coming of counter terminals and ATMs bank automation moved out of the back office and into the front office. Greater productivity and financial savings were not the only spur: banks were facing increased competition in services from each other and from other institutions.

Now, the twin forces of technology-push and financial deregulation are propelling the banking revolution forward with greater rapidity. Apart from reducing costs, technology is playing a key role in the provision of the new services needed to boost market share. It is also creating new competition for traditional bankers by enabling retailers like Sears Roebuck, for example, to become major forces in funds transfer. Checks have traditionally been used by banks to reinforce their dominance of the money transfer business. Now, deregulation and the massively increased use of plastic payment cards are opening things up. In the UK, the major banks are under assault from the building societies, and British banks and building societies are coming to resemble each other more closely. The banking profession is facing its greatest identity crisis.

Increased competition in turn increases the drive to keep down costs by investing in new technology to save on space and staff. New technology breaks down the barriers between traditional financial activities and is leading – through the introduction of "cashless shopping" by electronic funds transfer at the point-of-sale (EFT–POS) –

to the convergence of banking and retailing. Home banking and shopping also go together. The high-tech revolution is bringing about an *international* convergence of banking systems. It is speeding everything up and is shrinking the world by virtue of instant communications, making a reality of "fast bucks."

Keeping the Customer Satisfied

The face of retail banking is changing as increased competition forces banks to emphasize services to individuals. Cash dispensers and ATMs in "robot" banks, or "banking convenience centers," offering what is in effect self-service banking, are all part of the effort to keep the customers satisfied – and to stop them defecting to the competition. Banks will never be the same again.

There is general agreement as to what the bank of tomorrow will look like – the only question is how quickly it will arrive. Banks will be open-plan, with easy chairs and a more relaxed, informal atmosphere. ATMs in one section will relieve counter staff of the more boring and repetitive chores, freeing them to give financial advice and a more personal service tailored to each customer's needs. This will include home loans, insurance and travel requirements. Branch opening hours will be greatly extended, so customers can pop in at weekends and outside of normal office hours. ATM sections will be open 24 hours a day.

In Europe, recent reports from the EEC Commission and other bodies have sketched this futuristic scenario and the new-style banks have already started appearing in some parts of the Continent. Sweden is perhaps most advanced, with banks like the Götabanken in Gothenburg, while the UK's first fully automated bank – a TSB branch in Glasgow, Scotland – was opened in late 1985. Barclays and the National Westminster have experimental schemes up and running in Bristol, Basingstoke, Milton Keynes, Ipswich and Watford. Lloyds have a futuristic bank in Oxford Street, London.

However, as Catherine Smith points out in her book, *Retail Banking in the 1990s* (Lafferty, London, 1984), major manufacturers of banking hall equipment and the bankers themselves are still arguing over their different approaches to bank automation. People have also been talking about the "Bank of Tomorrow" for rather a long time now, and progress has been slower than expected.

The first high-tech gadgetry to appear in (or rather outside) bank branches were cash dispensers and ATMs. The dominant US manufacturer in the early 1970s was Docutel, which had 80 percent of the US market in 1976. But the market was small: less than 6000 ATMs had been installed in the USA and just under 5000 in Europe. Docutel has since given way to Diebold and IBM, who are now the undisputed US market leaders.

In the late 1970s and early 1980s, sales of the humble cash dispensers and ATMs grew rapidly. The USA had over 25,000 in place by 1982 and Japan had almost as many; according to the Battelle Institute, Europe trailed with about 10,000. By the end of 1985 the USA had 50,000 ATMs, Japan about 40,000 and Europe about 25,000, many of which were in France and Belgium. In fact, the largest ATM network in Europe is the 1600-machine system operated by the French bank, Credit Agricole. IBM and NCR are the European market leaders, followed by Dassault, Docutel, Transac and Chubb.

There is little doubt that ATM sales are set for steady growth, especially if prices continue to fall. Customers now seem to like them – indeed, there is UK evidence that people actually *prefer* dealing with these machines to interacting with human counter staff. The banks also like them, because getting the customer to key in the numbers means that someone else is doing the work normally done by bank staff. A lot of new technology shares this characteristic, and most of the major developments in commerce – like self-service supermarkets – have been built on the "labor-sharing" principle.

Many banks and financial institutions are now rushing to develop ATM networks. In the USA, systems like Cirrus, Plus, Visa's Electron and Mastercard's Mapp are growing rapidly, linking up smaller networks and giving Americans access to banking facilities outside of their home state. Visa's system uses a new type of debit card and is designed to pave the way for EFT-POS.

In the UK, shared ATM networks are also becoming important. In 1985, for example, a group of 21 banks and building societies led by the Funds Transfer Sharing consortium, the Abbey National and Nationwide Building Societies, the Co-op Bank and National Girobank launched Link, a shared network of 800 ATMs. The major British clearing banks such as Barclays and National Westminster have reciprocal arrangements for card holders. Citibank Savings, the British subsidiary of Citibank of New York and the major force behind Funds Transfer Sharing, is planning the largest investment in ATMs in Britain.

The very latest in ATMs are "talking" machines, which answer simple queries using recorded messages, and "video" ATMs, which show visuals of talking heads captured on videodisc. The tapes can be used to remind customers of the many different services the bank has to offer. In Europe NCR has demonstrated both kinds, but so far they have been slow to catch on.

Much in demand now are the new lines in counter terminals – individual workstations for counter staff – and branch processors, which are master terminals for managers. Such machines give the cashier and the manager instant access to information and transactions can be carried out in real time, rather than "batch mode," that is, waiting until the end of the day to cash up. Counter terminals speed up transactions, thus boosting branch efficiency and improving the level of customer service. This will hopefully enable some counter staff to spend more time with each customer, gradually transforming their role from routine cash-handler and administrator to that of personal financial adviser.

With the development of integrated database systems, bank staff in future will have a complete picture of every customer's financial affairs at their fingertips. This will enable banks to offer a more comprehensive banking and financial advice service and to build up a better relationship with their individual customers. Major banks are thus racing to develop "relationship" banking, and it is thought that those first in the field are likely to make a killing. "Integrated services" and "customer information" have thus become the key buzzwords and the name of the game among go-ahead bankers.

Bank of America, for example, has dramatically speeded up the processing of ATM and other electronic transactions – and cut the cost. Under Max D. Hopper, vice-president for technology, Bank of America has developed new software which allows bank staff to call up a complete profile of a customer's current account balances, loans outstanding and foreign exchange position just about anywhere in the world. This enables bank officers to know which services to push and gives them information on creditworthiness at the same time. Previously, such information was spread around the bank in about ten different systems.

Another important new customer service offered by the banks is electronic corporate cash management. This started in the USA in the 1970s as a response to the restrictions on US banks operating outside their own states imposed by the McFadden Act. What began as electronic balance checking to reduce information "float" developed into electronic cash management, that is, the placing in customers' premises of terminals that could tap directly into the banks' computers. There is currently a wide choice of systems – with fancy names like Chemlink, MARS, Bamtrac, Transend, Confirm and Cash Connector – on offer in both the USA and Europe.

Electronic cash management led, in turn, to the concept of automated treasury management, in which the corporate treasurer is provided with a very sophisticated "treasury workstation," which not only gives instant access to information but also has "decision support" software like the

Lotus packages built into it. Some argue that many large companies are now finding out that they can do a better job of managing their own money than the bankers themselves, and many small and medium-sized companies do not need such a system. But that hasn't stopped most major European banks climbing onto the electronic cash management bandwagon.

Electronic Banking: Fast Bucks

Electronic Funds Transfer (EFT)

Behind the scenes, banks have been steadily developing new ways of moving funds between each other more rapidly, using computers and electronics in place of messengers and paper. As long ago as 1918, US bankers and the Federal Reserve System established FedWire, which settles payments between banks that result from the totality of checks processed. Bankwire, managed by a consortium of US banks, is another system that has been in operation for many years. But such systems are slow, cumbersome and expensive.

The first major step toward automated clearing was taken in the UK with the establishment of the Bankers' Automated Clearing Service (BACS) in 1968 at Edgware in north London. Payments made through BACS are transferred directly into bank accounts from information stored on the users' magnetic tapes or disks which are taken to Edgware every evening from the City of London. In 1972, 166 million items were processed by the system, and by 1983 the annual total had reached 609 million. With user organizations growing at the rate of 25 percent a year, BACS could well be processing 2 billion items annually by 1990: this would represent about 40 percent of all UK interbank clearings.

Next came CHIPS (Clearing House Inter-bank Payment System), established by major New York banks in 1970. CHIPS replaced an earlier system whereby messengers carried bundles of paper checks to a central New York clearing house. As the volume of checks grew, the system was in danger of being swamped. But CHIPS easily coped with the 1970s check explosion, and the dollar value of checks processed has increased on average by 40 percent per year. CHIPS handles an estimated 90 percent of international interbank dollar transfers, and by the end of 1984 it was processing about 180,000 payments a day worth a total of nearly $200 billion. CHIPS has over 100 members already, and more banks are queueing up to join.

The first international automated clearing system was SWIFT (Society for Worldwide International Financial Telecommunication). SWIFT originated in Europe in 1977 and currently has its world headquarters in Brussels, Belgium. It transmits as many as 500,000 messages each day between more than 1500 banks in over 40 countries. The "old" SWIFT, or SWIFT I, had been doing so well in recent years – recording spectacular growth between 1980 and 1985 – that its centralized computer system came close to its physical capacity, and it was replaced by SWIFT II, which came on stream in 1986.

SWIFT II is being introduced country by country and should have completely replaced SWIFT I by the end of 1988. It is decentralized, it incorporates all the latest advances in technology, and it has a capacity of at least 1 million messages a day. President of the SWIFT consortium, Carl Reuterskiold, says that SWIFT also offers improved retrieval, better monitoring and control and a more sophisticated security system. Reuterskiold confidently expects to maintain current growth rates, but there is controversy over who can join SWIFT. Two international securities clearing systems, CEDEL and Euroclear, have been allowed to join, yet an application by stockbrokers was rejected in 1984.

The London version of New York's CHIPS is CHAPS (Clearing Houses Automated Payments System), which was finally introduced in 1984 after many delays and at least one major redesign. CHAPS allows large amounts (more than £10,000) of money to be transferred around the UK within seconds, thus making things like house and car-buying a lot easier. The new system was processing over 6000 transactions a day by the end of its first year of operation and looks set for steady growth.

Coming later than CHIPS, CHAPS has managed to avoid some of the former's pitfalls. It is less centralized, thus affording greater flexibility. It also uses the latest "non-stop" Tandem computer equipment, which offers greater security and reliability than earlier systems. Another major advantage of CHAPS over CHIPS is that the clearing banks guarantee the funds in the system – so there is no waiting until the end of the day to find out if the funds are bona fide. In future, CHAPS may well be extended to smaller, personal transactions (less than £10,000) and could pave the way for home banking. And there are plans afoot to integrate CHAPS with electronic corporate cash management systems.

But first, CHAPS has to overcome two major problems. The cut-off time for payments is 3 pm – in other words, half an hour earlier than the old manual town clearing system. The other problem is that banks other than the major clearing banks are not allowed direct access to CHAPS, but have to go through a clearer. The need to purchase extra interfacing equipment makes it very expensive for non-clearers, with the result that many foreign banks and merchant banks in the City of London have been refusing to use it. But once these difficulties are overcome, CHAPS will play a major role in the future City of London.

High-Tech on Wall Street

When the IRA bomb went off in the Grand Hotel, Brighton, in 1984, it is said that £10 billion stood poised to leave the City of London down a wire in the vital minutes before it was confirmed that Mrs Thatcher and the British Cabinet were safe. This was a dramatic illustration of the way in which huge sums of money can now be moved around the world electronically in a matter of seconds.

In Britain it is the "City Revolution" or "Big Bang"; in the USA it has been called the "Wiring of Wall Street." In each case, what is being described is the way financial institutions are being transformed by high technology. Much of what is going on remains unseen by the general public, yet it is of the utmost significance: the City of London accounted for one-quarter of the UK's entire foreign earnings in 1985. The new technology is eliminating paper, changing working methods, blurring the distinctions between traditional city professions, and increasing competition. Electronic markets are replacing the old "shouted" markets, shrinking the world and threatening the very existence of traditional financial centers like London, New York and Tokyo. For example, London's Baltic Exchange, a traditional world ship-broking center, is being challenged by a computerized ship-broking system based in tiny Bermuda.

The countdown to change began with the introduction of floating exchange rates in 1971, after years in which major Western currencies had fixed exchange rates. This was followed by an explosion of deregulation in financial markets around the world. In 1979, for example, the Thatcher government in Britain abolished exchange controls, and almost every other major country followed suit. But it was the new microchip technology just coming in that provided the means and the motivation for change. In a 24-hour business where speed is of the essence, going high-tech is no gimmick: it is a matter of life or death. It is in Wall Street rather than Detroit where you will find the very latest in high-speed computers and telecommunications technology.

Foreign exchange dealers and money brokers, spurred by the bonfire of foreign exchange controls, have rushed to install all-electronic dealing rooms. Floating exchange rates created the rapid increase in demand for financial information services, like Reuters Monitor and Telerate, which offer instantaneous information on currency movements throughout the world (see next section). Banks have turned to automation and new software packages to help keep track of their foreign currency positions. Last year in London, the Foreign Exchange and Currency Deposit Brokers Association launched its Automated Confirmation Service (ACS), which electronically links all London foreign exchange brokers to each other and to 130 banks. Previously, confirmations worth $80 billion on an average day were hand-delivered around the City by messengers. ACS was made possible only by new telecommunications technology introduced by British Telecom.

The formerly sedate world of stockbroking has also been transformed under the twin impact of new technology and deregulation, which reached a crescendo in London on October 27, 1986, when the "Big Bang" opened up the UK securities market by scrapping the distinction between jobbers and brokers and ending fixed commissions. The very survival of traditional stockbroking firms now depends largely on whether they get to grips with computers and computer-based services. Leading brokers are already using them for fund management, research and analysis and information distribution, while interactive networks are about to follow. Big Bang created a bonanza for computer companies as the City of London was transformed almost overnight into an electronic maze of computing systems (when they were working).

Stock exchanges themselves are changing beyond all recognition. In the London Stock Exchange, computers were first used to record settlements (the Talisman system, now being superceded by Taurus, a fully computerized system of logging changes in share ownership) and then to disseminate market information (the Topic service). In 1985, on its fifth anniversary, Topic was beefed-up to include the "real-time" prices of North American and European stocks. Topic now has 3000 terminals installed throughout the City of London and handles an average of over 800,000 requests for information every day. Foreign currency, traded option and money market information has been added to meet soaring demand for the service which has so far outstripped the original projections.

But even more significant is the fact that computers are now being used for the essential business of buying and selling shares. For this purpose, the London Stock Exchange is taking as its model the US National Association of Securities Dealers' Automated Quotation system (NASDAQ). This electronic over-the-counter market has increased its listing by 60 percent over the last ten years, compared with a fall of 1 percent on the New York Stock Exchange. NASDAQ's turnover rose six-fold over the period 1978–83. The London system is called Stock Exchange Automated Quotation (SEAQ), and it threatens to replace the trading floor of the London Stock Exchange with a network of video terminals dotted around in offices all over London. In 1986 SEAQ was linked directly to NASDAQ for the first time. But SEAQ already faces competition from Reuters, which has started a share dealing, as opposed to share information, system operated in conjunction with America's Instinet.

The real gleam in the eye of London Stock Exchange managers is the establishment of an integrated data network (IDN), first proposed in 1982. IDN will be a sophisticated and flexible electronic network, supporting a whole range of financial services and accessed by a single, simple terminal, such as an IBM PC. IDN will be developed in stages and will lay the foundations of a fully wired City. It is light-years away from the Dickensian world of clerks writing long-hand in ledgers, which characterized the City of London only a few decades ago.

Financial Information Goes On-Line

The last ten years have seen spectacular growth in electronic information services or "on-line databases." According to Link Resources, a New York research firm, in 1986 there were about 2400 databases available in the USA (up from 300 in 1978) and scores were being added every month. Spurred by the proliferation of personal computers and modems in homes and offices, the overall market for electronic information services was worth $1.5 billion by 1984 and was expected to reach $5 billion in 1987 and $10 billion in 1900, according to consultants International Resource Development. There is big money to be made from on-line services where it can be demonstrated that customers are willing to pay for information.

Thus, the boom in financial information services has been built on the insatiable demand in Wall Street for up-to-the-minute financial information. Stockbrokers, banks, insurance companies, foreign exchange dealers and commodity traders need instant information about their business on their screens in order to make instant decisions about buying and selling. No dealer can afford to be without the latest information tools. The financial information market may be somewhat elusive – though the on-line revolution has in a sense made what was once "invisible" literally more visible – but it's a market that was worth $500 million in 1985 and is expected to reach $3.4 billion in 1989. Small wonder that Telerate and Reuters Monitor are two of the fastest growing businesses in history.

The ending of fixed exchange rates in 1971 marked the starting point of the financial information bonanza. Successive deregulation measures have provided further boosts to the business. Quotron signed a contract with Merrill Lynch in 1972 to put equity stock quotations on-line. Reuters launched its comprehensive Monitor service in 1973. Meanwhile, a young man called Neil Hirsch started circulating prices for fixed interest stocks in a service he called Telerate. In 1972 Hirsch teamed up with Cantor Fizgerald to go on-line, and now the 39-year-old Hirsch finds himself president of a company that was valued at $880 million when it went public in 1983.

The financial information business can really be divided into three segments, each with its own leading players. First, there are firms supplying a basic service of stock quote prices, mostly delivered to brokerage firms. In the USA this sector is dominated by three firms: Quotron, which now claims 75,000 screens installed and 70 percent of the market, Automatic Data Processing (ADP), and Bunker Ramo. In the UK there is the London Stock Exchange's Topic service, started in 1980, plus a number of stockbrokers' in-house systems like Scrimgeour Kemp-Gee's Dogfox.

The second sector is the high rollers' table, with players who supply a broad range of on-line information from a variety of sources for a variety of markets. This market has become a classic duopoly featuring Reuters, whose Monitor system has 20,000 subscribers and 53,000 screens installed worldwide (but few in the USA), and Telerate, with 17,000 screens installed, mostly in the USA. Both Monitor and Telerate have grown spectacularly, with profits rising by 70–100 percent each year in the 1980s. Telerate – ironically, part-owned by the British company Exco between 1981 and 1985 – is rated higher on speed of delivery, though some say its technology is becoming out-dated. But subscribers to the London-based Monitor can deal in foreign exchange and bonds direct through the screen.

The third sector consists of more general "business information" services run by companies like Dun and Bradstreet, Dow Jones (which becomes AP/Dow Jones outside the USA) and Knight-Ridder. The Dow Jones business information package, called Dow Jones News/Retrieval, is a spin-off from the successful Dow Jones News service. Dun and Bradstreet has recently undertaken a major expansion of its DunsPrint and Who Owns Whom services in Europe with the purchase of Britain's Datastream and the opening of a £25 million database just outside London in 1985. In addition, there are more specialized business information services such as Manifest, which informs members of the London Commodity Exchange about the latest prices for gold, rubber, sugar, cocoa, pork, platinum and petroleum.

Reuters, the famous London news agency founded in 1850, made news itself in 1983 when it became apparent that its Monitor service had transformed the old firm's financial prospects. In fact, Reuters had turned out to be something of a gold mine for its owners, mostly loss-making Fleet Street newspapers. Eager to cash in on this unexpected bonanza, Reuters owners sold off 27 percent of the company in a share flotation in June 1984 which valued the company at £770 million (this was in fact an underestimate, for various reasons). The floating of Reuters sparked off a worldwide "treasure hunt" as it became apparent that "worthless" old Press Association shares purchased back in the 1890s were now worth millions. The Reuters gold rush was a remarkable tribute to the growth of the financial information business.

More recent developments include a tie-up between AT&T and Quotron to create a new Wall Street service and a joint venture between IBM and Merrill Lynch, who were planning to start a new service called Imnet. In 1985 Dow Jones, publishers of the *Wall Street Journal*, together with the Oklahoma Publishing Company, purchased the 52 percent stake in Telerate owned by Britain's Exco. Dow Jones plans a major expansion in the Telerate service.

Meanwhile, in yet another tie-up, Reuters has joined with Instinet to strengthen its position in the US market and to inaugurate a new interactive service that allows subscribers to deal in stocks, currencies and commodities worldwide via their screens. In 1985 Reuters also paid $58 million for Chicago-based Rich Inc., a supplier of communications systems for financial trading rooms. At the heart of the Rich system is a single console which can handle financial information from a variety of sources. This will help Reuters in the race to obtain the hardware for what has been called the "universal terminal" of the future.

The future of on-line technology lies with universal terminals, custom-built consoles, "intelligent" keyboards and monitors, expert systems to aid decision-making, and improvements in scanning technology which will make more information available (such as back-copies of newspapers) at the touch of a button. Whoever wins the hardware race – Reuters, AT&T, IBM/Merrill Lynch or Dow Jones – will have won an important battle in a market worth billions.

Paying with Plastic

Plastic cards – cash cards, check cards, credit cards, debit cards and phone cards – have revolutionized the business of making payments. Bank-sponsored credit cards first appeared in the USA in the 1950s, but

the real growth came in the late 1960s with the development of the Bank Americard and Master Charge national systems. The credit card was admirably suited to the USA, which lacked a nationwide banking system but had a well developed air transportation system and many frequent air travelers. Today, about 150 million people throughout the world carry plastic transaction cards and there is estimated to be over 800 million cards of one sort or another in use.

In Europe, which has national and linguistic barriers to contend with, plastic payment systems have recorded growth averaging 30 percent a year. Europe has three major payment systems: Visa (including Barclaycard and Carte Bleue), Eurocard (including Access and Master Card) and Eurocheque, the association of European banks. In Britain alone, Barclaycard and Access each boast over 8 million cardholders, while in Scandinavia great strides are being made with so-called universal payment cards like Sweden's Köpkort and Denmark's Dancard. In 1985 the European Council for Payment Systems initiated moves to make the three systems compatible, and post office banks from Japan and six European countries agreed to link their cash machines. These developments should make things much easier for businessmen, shoppers and tourists.

The use of plastic cards has undoubtedly cut costs and reduced some paperwork, but banks have been faced with the major and growing problem of fraud. For the one big drawback of plastic cards is that if they are lost or stolen there is no easy way of preventing their illegal use by a third party. Credit card fraud is costing the USA well in excess of $100 million and the UK £50–60 million per year. Some estimates have put the worldwide annual cost of card fraud as high as $800 million. The response of the banks and credit card companies has been to turn away from the standard magnetic stripe technology and to search for new, more secure systems.

The "ID Card," for instance, has a picture of the holder laminated into the card. US company Data Card International has produced a version that utilizes lasers to produce an image of the user's photograph made up of thousands of discrete picture elements (or "pixels") which are difficult to tamper with. Visa has come up with its "Electron Card" which embraces optical character recognition (OCR) codes and bar codes as well as a magnetic stripe. The Electron Card does away with embossed characters and those tiresome imprinted sales slips. Identix Inc. of Palo Alto, California, has developed a "smart" ID card which contains the digitized fingerprints of the user.

A multiple-feature card was adopted in the UK by the Committee of London Clearing Banks (CLCB) in 1984. The CLCB card has seven security features: (1) a flush signature panel, backed by (2) dyes which are released if the signature is tampered with, (3) special rainbow color printing, (4) fine background printing, (5) watermark magnetics, (6) hidden flourescent graphics and (7) a small thin film hologram. Other technical options on offer in Britain include a system developed by Rediffusion Computers and the National Physical Laboratory called "Signcheck," which analyses signatures using a ten-digit number, and Thorn–EMI's "Opticode," a signature-scrambling system that enables the counter assistant to read a scrambled signature with a special de-scrambling lens.

Holograms – three dimensional "photographs" of objects stamped onto thin sheets of material which are difficult to reproduce – are likely to feature more on credits cards in future. American Banknote has produced over 300 million postage-stamp-size holograms costing only a few cents each. This firm, and others like the small British company, Applied Holographics, are now trying to interest printers, publishers and packagers in the value of using holograms for a wide variety of purposes. For example, with the worldwide epidemic in counterfeiting, holograms can be used for the security labeling of products like fashion goods, jewellery, pharmaceuticals, computers and toys which are prone to being "pirated."

But most of the excitement recently has been caused by the arrival of the "smart" card invented by French journalist Roland Moreno in 1974. The "smart" card, or "carte a memoire," contains a memory chip that is "charged" with a certain amount of credit which can be spent at will. When the customer arrives at the point-of-sale, he/she keys in their personal identification number (PIN) and plugs in the card, which debits itself to the appropriate amount and records the transaction. All the relevant data is stored by the shopkeeper in a solid-state cassette called a "Cartette," which is then delivered to the local bank at the end of the day. This system seemed particularly appropriate for France, where checks are costly to process and the phone service is somewhat unreliable.

"Smart" cards got off to a slow start, although there were successful local experiments at Caen, Lyons and Blois. Indeed, some wrote them off as an expensive French novelty. But Moreno persisted, first signing up the French PTT, which wanted them for use in conjunction with its ambitious Minitel videotex system. Carte Bleue, Visa's French affiliate, also began test-marketing a "smart" card. In 1986 the French bank organization, Carte Bancaire, ordered no fewer than 16 million "smart" cards – 12.4 million from Bull and 4.2 million from a Philips subsidiary – to equip every French credit card holder.

Originally, there were four companies making "smart" cards: Honeywell–Bull, the French–American company; Philips of Holland; French-owned Schlumberger; and Smartcard International of New York. But now the Japanese, in the shape of Casio, Mitsubishi and Dai Nippon, are making a big play for the "smart" card market and IBM is expected to join at any time. Master Card has been testing 25,000 Honeywell-Bull cards in Columbia, Maryland, and is running a similar experiment with 15,000 Casio cards in the Palm Beach area of Florida. Visa, American Express, Carte Blanche and Diner's Club are also believed to be carrying out studies and secret "smart" card trials. The US Agriculture Department has even supplied 150 "smart" cards to Georgia peanut farmers. But despite the snowballing enthusiasm, not everyone is convinced that "smart" cards will take off in a big way. They may, for example, be used mainly as a marketing gimmick among upscale customers.

"Smart" cards have other uses, such as controlling computer access to databases and physical access to missile bases. For instance, the US Department of Defense has used a Philips "smart" card to control entry to its personnel policy office at Fort Lee, Virginia. "Smart" cards have also been used to improve the security of banking messages and in a home banking experiment at Fargo, North Dakota. Some have suggested that they be used to record personal medical histories, to record transactions on the floor of the New York Stock Exchange and to stop food stamp fraud. Micro Card Technologies Inc., the Dallas-based subsidiary of France's Bull, is bullish about the future for "smart" cards and says it plans to start production soon in the USA. Casio's Microcard Corporation is equally confident about future sales, although admitting that a lot depends on whether mass-production can bring down the cost of each card.

Apart from "smart" cards and cards incorporating holograms, the future of plastic payments systems might lie with "laser" cards and "bioaccess" systems. For example, the Drexler Corporation's LaserCard can store vast amounts of information thanks to a technique that uses lasers to insert and read "pits" in a protected layer of silver particles suspended in gelatine inside the card. It's all part of the optical storage revolution that's already brought us the videodisc and the CD ROM. Bioaccess systems scrutinize parts of the card user's body – fingerprints, palm geometry, the voice and eye retinas – to provide perfect security. Retinal scanning (of the pattern of blood vessels in the eye) is claimed to be the most reliable, but it will be some years before such systems become cost-effective. In a recent James Bond movie, a villain manages to fool just such a system.

The Retail Revolution

Bar Coding

In the late 1970s, predictions were made of a revolution in retailing arising out of the use of electronic point-of-sale (EPOS) equipment. But they proved over-optimistic: in Europe especially, progress with the installation of EPOS terminals incorporating laser scanners to "read" the new bar codes on everyday products has been slower than expected, owing to the combined impact of recession, high costs and a conservative "wait-and-see" approach among many retailers. But now there are clear signs that things are at last beginning to happen. There is also no doubt that computers will play an increasing role in many aspects of retailing in the years ahead. The Japanese, for example, have the futuristic Seiyu Nokendia store in Yokohama, which features robot unloaders, stackers and transporters and liquid crystal price tags, as well as EPOS terminals.

Bar codes first started appearing on grocery products after the US

Grocery Industry Ad Hoc Committee adopted the Universal Product Code (UPC) in 1973. Each product has its own code, which consists of a series of black and white bars printed on the product package. The code (in Europe it is called an EAN, or European Article Number) identifies the product's manufacturer and its size and weight as well as its price. The principle behind the UPC is simple: each code is made up of binary digits – a narrow bar might represent "0" and a wide bar "1." To read the code, a laser scanner shines a light across the bars and then transforms the resultant patterns of light into electrical pulses.

When the checkout operator passes the printed bar code on each item over the laser scanner built into the EPOS terminal, the code is "read" and the information transmitted to the store's computer. Instantaneously, the current price for the item is fed back to the checkout where it is displayed, together with a description of the item, on the VDT at the checkout. At the same time, a printout of the product and price is provided on the till receipt. All this takes place in a fraction of a second.

Thus, one advantage of laser scanning is that the checkout process can be speeded up, because the checkout operator doesn't have to key in the price or a string of code numbers for each item. Fewer checkout operators are therefore needed for a given number of customers, and this leads inevitably to a reduction in labor and labor costs. The store can change prices instantly without having to relabel every can or box, simply by entering the alteration on the central computer. Laser scanning also allows close monitoring of checkout performance and eliminates checkout errors such as "under rings," cuts down on "shrinkage" and provides management with the means for instant stock control: at any time, the store's boss can find out what lines are selling and how quickly, just by tapping into the information supplied from the checkouts to the central computer. Instructions to the stockroom, warehouse or outside suppliers can swiftly follow.

But consumer associations and labor unions see disadvantages in laser scanning and bar codes. The removal of price tags on each item makes "comparison shopping" more difficult and could lead unscrupulous store-owners to raise prices surreptitiously. Any reduction in "price consciousness" could work against the consumer and shift the balance in favor of the retailers. (A joint economic committee of Congress found that US supermarkets had overcharged customers by $660 million in 1974.) In many US states, stores have been forced to reinstate price tags – which of course reduces the cost benefits of the system. In the UK, however, a special Office of Fair Trading report on the impact of new technology in retailing concluded in 1982 that no new laws to protect the consumer were necessary.

Labor unions have been concerned chiefly about the job implications of scanning and see its introduction as part of the historical drive to reduce manning levels in retailing. The biggest retail innovation this century – the invention of supermarkets in the USA in the 1930s – *was* specifically designed to reduce retail labor: as one early supermarket boss put it at the time, "We aim to make the shopper the store's best employee." Scanning has not resulted in anything like as many jobs being lost, and most reductions in the number of checkout operators have been achieved painlessly through normal turnover and the relocation of labor within stores.

Bar codes have many uses outside retailing, and indeed it is in other areas that the greatest growth in their usage is now being experienced. They are being used wherever there is a need to track inventory – on assembly lines, in warehouses, in the docks, on ships, in hospitals, in libraries and even among runners in the New York City Marathon! The US Department of Defense went over to bar codes for all its supplies in 1982. Giant Food Inc. has a huge state-of-the-art automated warehouse complex at Jessup, Maryland, which uses thousands of scanners to speed the flow of goods to Giant's stores.

In the past, such inventory checking was done by employees armed with pencils, clipboards and sometimes bicycles. Now, the wave of a wand scanner or the flash of a low-power laser beam enables manufacturers and distributors to know exactly how much has been produced and where everything is. Industrial coding systems, pioneered in the USA by market leader Intermec Corporation, have experienced a sales growth of 40–60 percent in recent years and the market is expected to be worth $1 billion by 1990 (up from $170 million in 1984). The only rivals to bar codes are optical character recognition (OCR) systems and electronic tags that transmit information by radio signals.

While the USA had about 13,000 scanning stores in 1986 and Japan about 5000, progress in Europe and the UK has been a lot slower. By the end of 1986 the UK had a mere 350 scanning stores, and the total is expected to reach 1000 only by the end of 1988. Despite some successful early experiments, predictions made in the early 1980s about one in four UK stores going over to scanning by 1984 just didn't materialize. Even today, the experts disagree as to whether or not this aspect of the retail revolution is about to happen: recent reports from UK computer makers ICL and analysts Post News predict an explosive growth in sales of EPOS terminals, while consultants Euromonitor and the European Foundation for the Improvement of Living and Working Conditions have highlighted the continued reluctance of major retailers to invest in laser scanning systems.

Tesco, an early UK enthusiast, had only a handful of scanning stores in 1986, while J. Sainsbury led the grocery chains with its total of scanning stores – just 34. However, stationer W. H. Smith is pushing ahead with the installation of EPOS terminals following a successful experiment with a sophisticated new management information system. With 70,000 individual stock items and 360 major stores, W. H. Smith found it was spending huge sums of money on stock checking and handling. Now, light pens at the checkout installed by National Semiconductor are helping make the firm's operations much more efficient.

High technology is not only transforming the physical handling of goods by retailers and wholesale distributors: it is also transforming the work of salespeople, advertisers and market researchers. For example, automobile sellers are now turning to high-tech marketing in the form of videotex and videodisc systems which are used to answer technical questions, to show potential customers the range of models available and to track down the whereabouts of cars with particular specifications. Said one sales manager, "Today's consumer is just too sharp, too well educated and too on top of technology to fall for the blue-suede-shoe salesman anymore."

Traveling salespeople are now going out on the road equipped with portable personal computers and videodisc players. EPOS terminals are being hooked up in sophisticated market research experiments by firms like IRI, A. C. Nielsen and BehaviorScan, in order to pinpoint which families are buying which products in response to which TV commercials. All-electronic retailing has also arrived – in the shape of CompuSave's free-standing videodisc kiosks which are appearing in some US stores. But so far they have provided little threat to people like Comp-U-Card, who have a 60,000-item catalogue available.

Cashless Shopping or EFT/POS

The logical solution to the problem of handling cash, checks and multiple credit card forms at the checkout till is direct electronic funds transfer at the point of sale (EFT/POS) – so-called "cashless shopping." Since the majority of consumer financial transactions occur in stores, the advantages of EFT/POS to the banks is obvious: it would eliminate a great deal of paper. With EFT/POS, the shopper inserts a debit card into the checkout terminal and enters his or her secret personal identification number (PIN); the checkout assistant enters the amount on the keyboard, and the customer's bank account is instantaneously checked and debited over the phone line. Actually, there are four types of

EFT/POS system: the "on-line", the "off-line" (which debits accounts only at the end of the day), a combination of the two (on-line credit-checking, day's-end debiting), and the "smart" card already discussed, which is emerging as a competitor to EFT/POS systems in its own right.

Unfortunately, the tantalizing dream of smooth, cashless shopping has not been matched by real achievements, and there is no country in the world that can claim a major, successful EFT/POS system in operation. There are many pilot schemes – such as the Limoges project in France and others in Denmark, Norway, Holland, Belgium and Austria. But EFT/POS in general has yet to get off the ground. In the UK it has become a rather ridiculous on/off saga. Various local experiments have been conducted in Nottingham, Wilmslow, Tunbridge Wells, north London and the Thames Valley. Specialist retailers (Marks & Spencer, House of Fraser, etc.) have launched their own schemes. But the Committee of London Clearing Banks' involvement in a proposed national EFT/POS network has resulted in farce.

The CLCB first set up a committee to study the implementation of EFT/POS as far back as 1978, but a subsequent scheme in Southampton was hastily abandoned and a new study team appointed in late 1981. That one reported in May 1983, stating that an EFT/POS pilot scheme using IBM and British Telecom equipment would go "live" in 1986. But there had been insufficient consultation with the retailers that would actually be using the stuff. Consumer groups were upset by the way the CLCB was making its plans in secret. Most important of all, the question "Who Pays?" had not been answered by the CLCB committee – would it be the banks, the retailers or the consumer?

The retailers could see few advantages for themselves, and meanwhile some credit card companies were making their own arrangements with transaction telephones and the like on a piecemeal basis. To cut a long story short, the CLCB, in July 1984, announced that its pilot scheme was being postponed while different systems from different manufacturers were evaluated. But in 1985 the CLCB said the national EFT/POS scheme was being revived for 1988, when three pilot projects would be run.

Home Banking and Shopping

Potentially even more revolutionary than nationwide EFT/POS would be the widespread use of TVs, telephones and personal computers for the purposes of home banking and shopping. Once again, early predictions about the rapid spread of home banking and shopping have not been realized. Despite a number of fairly successful experiments in the USA and Europe, home banking and shopping is still essentially at the experimental stage. But greatly increased sales of personal computers for home use could give banking and shopping à la modem a big boost.

To take advantage of home banking facilities, customers need a personal computer or similar device and a modem that enables them to hook up via the telephone line to their bank's computer. After punching in their PIN, customers can then display their current balances and account details on their TV or personal computer screen, pay bills, transfer money from one account to another, call up statements, order checkbooks and ask the bank questions. Home shopping or "teleshopping" systems work in much the same way. After the customer has viewed the goods and made a selection, payment is authorized by tapping in a credit card number.

In the USA teleshopping got off to a slow start, although Warner-Amex's famous Qube experiment in Columbus, Ohio, showed that it was feasible. In 1983 Knight-Ridder launched its Viewtron experiment in south Florida using AT&T's Sceptre terminals, but takeup was slow because of the expense and it was later abandoned. Times-Mirror, Field Enterprises and an IBM–Sears–CBS consortium set up similar experimental schemes using personal computers rather than special terminals, but participation was limited, and they too were axed.

A recent Yankee Group study of teleshopping in the USA identified certain critical factors holding back progress: in particular, shopping from home did not meet the *psychological* needs of most shoppers, who were keen to handle the high-value items in particular and enjoyed the social aspect of going out to shop. They were also worried about delivery times, payments systems and after-sales service. The Yankee Group concluded that home retailing is unlikely to be big until at least the 1990s.

On the home banking front, Bank of America's San Francisco scheme had attracted 15000 subscribers by 1985 and Chemical Bank's Pronto service claimed 21,000 users. But these totals were a far cry from the huge numbers originally envisaged (Chemical Bank has 1.15 million customers). Some banks – like the Los Angeles First Interstate, San Francisco's Crocker National and Miami's Dadeland Bank – have quietly dropped their home banking services and/or plans altogether.

Although Bank of America and Chemical Bank (in conjunction with AT&T and Time Inc.) are adding stock brokerage services to their systems, customers are not exactly clamoring to go on-line; they are clearly apprehensive about using high technology for private transactions and fear security problems. But home banking systems also have two very serious (and rather obvious) drawbacks: first, they don't allow customers to withdraw cash or make deposits – for these transactions you still have to trek to the local bank branch; and second, most domestic customers do not carry out enough transactions to justify the high monthly fees.

Similar conclusions have been drawn from the half-dozen or so home banking and/or shopping experiments conducted in the UK in recent years. One of the first, the Gateshead Shopping and Information Service, started by Tesco and the University of Newcastle in 1979, found that home shopping was very useful to disadvantaged groups like the elderly, the housebound and the disabled, but that most other people saw shopping as a great opportunity to get *out* of the house.

Homelink, launched by the Nottingham Building Society and the Bank of Scotland using free Prestel adaptors in 1982, was a modest success: more than half the subscribers used the service several times a week – although mostly for looking up their bank and building society balances. In 1984 stockbrokers Hoare Govett began a home brokerage service, also via Prestel, the British Telecom videotex service. Out of the Homelink experiment in 1985 came the Bank of Scotland's fully fledged Home Banking service; this also worked via Prestel and was the first such service from a major British bank.

Club 403 was a service started by the Birmingham Post and Mail Group's Viewtel subsidiary in the West Midlands in 1982. Backed by the UK government in an attempt to find a domestic market for UK videotex technology, it offered betting and booking services, video games and educational programs as well as home banking and shopping. But the reception was mixed, and the potential armchair shoppers and bankers were easily put off by complicated instructions and lack of choice. A majority were also unwilling to pay very much for the privilege of such home comforts.

In 1985 Britain's Littlewoods Organization – a major football pools, mail order and retail group – launched its Shop TV, a service enabling 50,000 Prestel users to purchase a range of up to 300 branded domestic electric goods, such as VCRs, microwave ovens, refrigerators and freezers, at discount prices. The similar Telecard service was launched in London. This is business that retailers will not want to lose. Shop TV represented the first time a major retailer had offered a teleshopping service, although many experts consider that the size of the UK market would remain quite small for some time.

Most bankers say that home banking is well down their list of priorities, while most retailers are quite happy to adopt a "wait-and-see" approach to home shopping. At present, consumers do not appear to be clamoring for either. And in the final analysis, it is the consumers – not the equipment manufacturers, not the banks, not the retailers and still less governments – who will decide which new financial and retail services meet their real needs. That is a major lesson of the story so far of micros and money.

9 Key Problems for High-Tech Society

High-Tech, Smokestacks or Services: Where Will the Jobs Come From? – What Kind of Work? – Managing Technological Change: Will Labor Be Left Out in the Cold? – High-Tech Crime: Are Hackers Heroes? – The Threat to Privacy

The emerging high-tech society will not be without its problems. Problem number one is the future of work – the quantity of work, the quality of work, and the organization of work.

When the magical microchip burst onto the public scene in the late 1970s, wild predictions were made about its likely impact on employment. In the UK, books like Chris Evans's *The Mighty Micro* (Gollancz, London, 1979) and Clive Jenkins and Barrie Sherman's *The Collapse of Work* (Eyre Methuen, London, 1979) attracted much attention with their dire predictions of mass joblessness.

Two major reports – one for the British government by Iann Barron and Ray Curnow (published as *The Future With Microelectronics*, Frances Pinter, London, 1979), the other for the French government by Simon Nora and Alain Minc (published as *The Computerization of Society*, MIT Press, Cambridge, Mass., 1980) – argued that the microchip would soon become so pervasive that jobs would be replaced by computers on a massive scale.

In the context of rapidly rising unemployment in Europe, some of the more fanciful projections – for example, the prediction in a report to the West German electronics company Siemens that "40 percent of office jobs will go by 1985," or that only a fraction of the current workforce would be required to do "all the work" by the turn of the century (Professor Tom Stonier) – actually gained credibility.

At the same time, the 24-nation Organization for Economic Cooperation and Development (OECD) reported that some 20 million new jobs would have to be created by the end of the 1980s just to hold unemployment stable at 35 million in OECD countries. It's hardly surprising that such revelations created a major scare about the chip, which was even dubbed by some the "job destroyer." The renewed speculation also sparked a new debate about automation, the future of work and the role of the "work ethic" itself.

Unrestrained chip enthusiasts and techno-boosters made wildly optimistic claims about the job-creating potential of high-tech. The only employment problems, they claimed, would be short-term, transitional ones. Looking on the bright side, they repeated the familiar, orthodox argument that new technology means higher productivity, which in turn means greater wealth and thus more jobs and prosperity for all in the long run. It was not automation itself but the *failure* to automate that risked jobs, said the optimists. The problems of structural change had been faced and overcome before. Jobs had been created in large numbers for a growing labor force in the past and this would happen again. After all, hadn't the Luddites of 1811 been proved wrong when employment in the British textile industry continued to grow during the nineteenth century?

By the mid-1980s, it was evident that neither the pessimistic Cassandras nor the optimistic Pollyannas had won the employment argument. The onward march of the microchip had been fairly slow. A host of technical, practical and financial problems – not least of which was the persistence of the economic recession itself – had reduced the employment impact of new technology. There had been no overnight Armageddon and no mass takeover by robots and computers.

Yet it was also obvious that things would never be the same again. The shakeout in traditional manufacturing industry had been severe – more severe than almost everyone had predicted. The *pattern* of employment was changing swiftly and significantly with the switch to the service industries. Even the optimists had to admit that there would be a "jobs gap" between the supply of and the demand for labor in most advanced industrial nations for the foreseeable future.

There were also dire warnings for management and labor, who were both struggling to cope with technological change. Information technology was transforming managers' jobs, restructuring individual companies and turning whole industries upside down. Labor unions had to begin a long rearguard action designed to ensure their very survival in the coming high-tech society, as collectivism was supplanted by a new individualism.

A reported increase in computer frauds and some well-publicized examples of "hacking" (gaining unauthorized access to computers) led to widespread concern about computer security and fears of a high-tech crime wave. Computers themselves became increasingly useful in the fight *against* crime, along with other new devices, but they also posed a threat to privacy as cheap computing power enabled private and public organizations to store on computer more and more information about individuals. A whole battery of new high-tech surveillance techniques came on the market. But it was the future of employment that became the main worry.

High-Tech, Smokestacks or Services: Where Will the Jobs Come From?

While the professional pessimists' dire predictions of massive technological unemployment have failed to materialize, there is no doubt that high technology *is* reducing the demand for labor in all sorts of areas and in all kinds of ways. There may have been no "collapse" of work, but there has been a steady erosion of employment opportunities. Leaving aside some of the more abstruse theoretical arguments and philosophical speculations in order to concentrate on what is actually happening, we can identify four major causes of concern.

First, high-tech industry itself is low on jobs. Because the high-tech sector is small, and productivity is rising in line with output, it can only ever create a modest number of jobs. Second, the savagery of the shakeout in the smokestack industries like steel and automobiles has been much greater than expected, and it is also clear that those millions of Rust Bowl jobs will never come back.

Third, despite the efforts of the Great American Job Machine, which has created 20 million jobs in the service industries in ten years, there are

now doubts about the service sector's continuing ability to generate jobs in large numbers. And fourth, while it is generally accepted that manufacturing will still play a key role both in national economies and in the occupational structure, there is a growing realization that any expansion of domestic manufacturing capable of meeting foreign competition is unlikely to create many new jobs because of the necessary adoption of automated methods – this is the so-called phenomenon of "jobless growth."

High-Tech Is Low On Jobs

Studies of the job-generating potential of high-tech suggest that there is little warmth in the so-called "sunrise" industries. The majority of the jobs of the future will in fact be low-tech – in such humble occupations as cashier, caretaker and secretary (see figure 9.1, p. 252). According to Bureau of Labor Statistics and Data Resources Inc. figures, high-tech industries in the USA (generously defined) will generate only between 750,000 and 1 million jobs in the whole country in the next ten years – this is less than half of the 2 million jobs lost from US manufacturing industry in the period 1980–3 alone. And of these new jobs, fewer than a third will be for engineers and technicians – the rest will be for operatives, clerical workers and managers.

The total number of jobs that will be created in high-tech is fairly modest because the high-tech sector is relatively small. The high-tech industries employ a mere 3 percent of the US non-agricultural workforce – and this will climb only to about 4 percent by 1995. The output of the entire high-tech sector is less than twice that of the automobile industry. Rapidly rising productivity – outstripping that of other industries – will further depress the demand for labor as high-tech firms automate their own production lines. In addition, unskilled jobs in high-tech, such as those of assembly line operatives, are always under threat from low-cost foreign labor. Companies can easily export routine production or subcontract work overseas.

A Stanford University study by Henry M. Levin and Russell W. Rumberger (*Technological Forecasting and Social Change*, vol 27, no. 4, 1985) concluded that high-tech was being oversold as a source of new jobs. While employment in jobs related to high-tech would jump 46 percent by 1995, this would account for no more than 6 percent of all new jobs created in the economy. Not one of the occupations in the growth Top Ten was related to high-tech: "Technology has a place – but by no means a dominant one – in the job market of the future," according to Levin and Rumberger.

Confirmation of the fragility of high-tech employment came with the Great Silicon Valley Shakeout in the summer of 1985. A series of plant closures, mass lay-offs and extended vacations rocked the Valley, leading one observer to comment, "The party is over." In June, Apple Computer made an astonishing 21 percent of its employees redundant in one fell swoop, and even Intel cut its payroll by 11 percent. That meant that total employment in the US semiconductor and computer industries had fallen an astounding 27 percent over the first half of 1985, to just 736,000. The slump in employment was not confined to computers: in August 1985 telecoms giant AT&T announced that 24,000 out of its total workforce of 117,000 persons (about 20 percent) would go by the end of the year.

And it was the same story in Europe: the major West German electronics company, Siemens, was steadily reducing its workforce, while Britain's only mainframe maker, ICL, had seen its payroll decline from 34,000 in 1980 to less than 15,000 in mid-1985. A 1985 report from consultants Cambridge Econometrics concluded that employment in the UK computers, electronics and telecoms industries was likely to decline by about a fifth by 1990, despite a steady rise in output.

Smokestacks Won't Come Back

While high-tech is creating jobs in only modest numbers, traditional smokestack industry continues to shake out vast numbers of employees. There is a growing realization that these jobs will never come back. Although total manufacturing employment (including high-tech) in the USA will actually grow from today's 19 million to 22 million by 1995, according to Labor Department figures, employment in the old Rust Bowl industries such as metals, rubber and chemicals will shrink from a huge 20 percent of the total US labor force to a mere 8 percent. Meanwhile, employment in services will leap from today's 20 million to nearly 29 million by 1995.

An archetypal example is the US automobile industry based around Detroit, Michigan. From 490,000 workers in 1958, employment in car and supplier industries rose to a peak in 1978 of 941,000, of whom 580,000 were directly engaged in building cars (output tripled over the same 20-year period). But in 1982–3, no less than 500,000 employees in the wider industry suffered lay-offs of varying lengths, and by 1986 it was apparent that many of the lay-offs had become permanent. It is now widely anticipated that total automobile industry employment will be down 20 percent by 1990, while output remains about the same. Detroit plans to install 25,000 robots by the early 1990s, and these alone could displace 75,000–100,000 workers. As a footnote on the shrinking power of labor unions, the United Auto Workers (UAW) has lost 400,000 members since 1978.

All over the Rust Bowl region of Michigan, Indiana, Ohio and Pennsylvania, where unemployment is well above the US average, the jobs story is the same: in the blighted steel-making "Mon Valley" on the banks of the Monongahela River near Pittsburgh, US Steel has reduced its workforce from 28,000 in 1979 to less than 9000 today. In the wider Pittsburgh area, jobs in steel have plummeted from 90,000 to 40,000 over the same period. In the Rust Bowl as a whole, 230,000 out of 577,000 steel jobs have gone for good, apart from the thousands more in other metalworking and appliance manufacturing trades. And it's not just the manual workers who are suffering: a recent survey suggested that over 1 million white-collar workers have lost their jobs in smokestack industries, including 40 percent of those employed in the steel industry.

Birmingham, the British equivalent of America's "Motown," has also suffered appallingly. A report published in 1985 by Britain's Economic and Social Research Council showed that the city's industrial base shrank by a third in the 1970s. The 26 largest companies cut their workforce by 40 percent as the key sector of cars declined. In Britain's major cities, employment has fallen 45 percent in the last 30 years.

A study for the British government by researchers at the Universities of Aston and Sussex found that over 1 million jobs were lost from UK manufacturing in 1980–2 alone, compared with 1.5 million over the ten years from 1970 to 1980. Vehicles, mechanical engineering, textiles and clothing suffered declines in 1980–2 of nearly 20 percent. A more detailed study of the West Midlands, Britain's industrial heartland, by the same researchers concluded that it will be many years before the employment gains from new high-tech industries can begin to offset the enormous losses from the smokestack industries. A 1986 Policy Studies Institute study found that job losses due to the introduction of new technology in traditional UK manufacturing were increasing.

If there are any remaining doubts that jobs in traditional UK manufacturing won't return, they are easily dispelled by a close monitoring of developments in sectors as diverse as textiles, paper, rubber, food and drink, coal mining and consumer goods. Labor-shedding is the order of the day, as UK companies struggle to meet foreign competition. As one industry spokesman, Peter Caudle of the Chemical Industries Association, put it in 1984, "The whole ethos now, I'm afraid, is to boost production without bringing back jobs which have been most expensively shed." Doug Hall of Hall Automation, Britain's only robot maker, confirmed that "Many large companies are pursuing a policy of de-manning"; While former Austin Rover chairman, Harold Musgrove, stated: "The day is not far off when entire areas of our plants will run automatically with minimal human supervision night and day, seven days a week."

A New Service Economy?

In the mid-1980s, the Great American Job Machine became the envy of the world. Unemployment in Europe was rising to record levels, but in the USA it was falling. Between January 1983 and June 1984, the US economy created 4 million new jobs, mostly in the service industries. Over the period 1973–83, the USA had gained 13.2 million employees, while Europe had suffered a net loss of 2.9 million. In the first two months of 1984, the USA had created more jobs than the EEC had generated in the previous 14 years. By December 1985 US unemployment reached a five-year low of 6.9 percent of the workforce. Meanwhile, in Europe there were signs of recovery, with the OECD predicting a net gain of 800,000 jobs across Europe in 1986 – the best increase since 1979.

Of the total US workforce of around 105 million, roughly two-thirds were now in the service sector. Between the end of 1979 and the end of 1984, service sector employment rose by 8.6 percent, while employment in the production of goods fell by 5.5 percent. The largest increases came in business services (up 35 percent), health (up 19 percent), eating and drinking (up 14 percent) and finance, insurance and real estate (up 12 percent). The eating and drinking category – now accounting for over 5 million US jobs – of course includes the fast-growing fast-food sector: this prompted one British newspaper to refer to the "Great American Jobs-Burger."

There seem to be half a dozen basic reasons why the USA has been able to generate so many jobs. First, a rising US population (14 percent up between 1973 and 1983) helped to increase labor demand as well as supply, because each extra person has needs that take another worker to fill. Second, GDP has been growing – thanks in part to greater productivity in farming and manufacturing – with the result that people have been consuming more services.

Third, slow productivity growth in services (services have fewer economies of scale) has meant that service jobs have increased roughly in line with the increased demand for services. Fourth, foreign competition is not so much of a problem in services – you can't import the services of a hairdresser – so increased demand again feeds directly into domestic employment creation.

Fifth, a high proportion of new service sector jobs have been generated by small business. The American genius for entrepreneurship – spotting business opportunities, figuring out new ways of making money – has come to the fore and stands in stark contrast to European complacency and pessimism. According to Professor David Birch of MIT, small companies in the USA with less than 500 employees created no less than 2.7 million jobs between 1980 and 1983. In California they account for 70 percent of all new job growth. Even in Rust Bowl states, an impressive number of service jobs have been created by young entrepreneurs whose average age is 33 years, according to one survey.

Finally, American flexibility – in contrast to European rigidity – has enabled the US labor market to adapt more rapidly to the new economy. Wage rates have gone down as well as up – real wages fell an average 11 percent during the 1980–2 recession – with some industries, like airlines, recording massive pay cuts. This has enabled companies to retrain or recruit more labor after temporary lay-offs: in heavily unionized Europe, where wage-cuts and lay-offs are much less common, companies have preferred to substitute capital for labor wherever possible. Labor mobility is also much greater in the USA – no less than one in six Americans now moves house each year. Most Europeans, on the other hand, are much less willing or able to move in search of work.

However, even in the USA, doubts are beginning to creep in about the continuing capacity of the service sector to generate so many new jobs. As early as 1982, both the General Accounting Office (GAO) and the Bureau of Labor Statistics (BLS) questioned whether office and service sector employment would still expand once new technology began to penetrate in a big way. In another study, Charles Jonscher of MIT predicted that jobs in the "information" sector would not increase as rapidly as they had done in the past. It was also clear that the massive new investments in technology being made in service industries such as the public utilities, transportation, distribution, hospitals and hotels was bound to boost service sector productivity.

In the long run, this will lead to pressure on service sector jobs. "What we are seeing," wrote Irving F. Leveson of the Hudson Institute, "is the rapid industrialization of the service sector." Investment in new technology meant that the service sector was rapidly centralizing its operations, moving production facilities to lower-cost regions of the country, standardizing its output and finding new ways to assemble, sell and distribute its goods more efficiently. The American practice of franchising was making possible the mass production of services.

Economists couldn't seem to agree whether rising service sector productivity would be good or bad for employment. By 1985 some were arguing that these more productive, better-paid service sector jobs coming along would generate economic growth and thus more jobs in the long run. Others argued that the average service sector job would still pay only about 70 percent of the average in manufacturing for the foreseeable future, and service sector productivity would remain lower than that in manufacturing for a long time to come, thus inhibiting economic growth. Most seemed to accept that the service sector couldn't go on expanding ad infinitum without a strong, underlying industrial base. As industrial policy advocate and MIT economist, Lester C. Thurow, put it, "If Industrial America is dying out, then those services used by industry will die out, too."

In Britain, where manufacturing employment was down to a rump 26 percent of all jobs in 1985, there was new evidence of a declining rate of job creation in the service sector. For example, the increase in bank staff had slowed to a virtual standstill by 1985, despite an increase in payment items. Various studies were predicting a steady decline in banking employment – although they disagreed as to whether it would set in before or after 1990. An Institute of Manpower Studies report foresaw stable or shrinking numbers employed in clerical, secretarial and related occupations, which alone account for one-sixth of the UK workforce.

While doubts were being raised about the *quantity* of jobs being created in the service sector, there was also growing criticism of the *quality* of many of the new service jobs. Service sector jobs are less well paid and offer few fringe benefits. A huge proportion are in fact part-time jobs, and two-thirds of the service jobs created in the USA since 1979 have been taken by women. Apart from low pay, part-time workers typically get no health care, life insurance or pension. They have little job security and are more than likely to be used as pools of temporary labor to be taken on and laid off as demand dictates.

This trend to voluntary (and involuntary) part-time work is creating a new group of second-class citizens, which now accounts for nearly 20 percent of the US labor force. Some employers have welcomed the trend to part-time working, seeing it as an easy way to cut labor costs in the face of foreign competition. Labor leaders are fighting the trend by resisting the growing efforts to replace full-time jobs with part-timers. They see it as a symptom of comparatively high unemployment and warn that it could lead to a workforce divided between those in secure, skilled and well-paid positions and a larger stratum of unskilled workers drifting in and out of poorly paid, temporary jobs.

Jobless Growth

In response to the growing doubts about the quantity and quality of the new service sector jobs, many economists, politicians and businessmen have come forward to argue that manufacturing still has a major – indeed, the key – role to play in national economies. Every country needs a strong, underlying industrial base because manufactures are more tradable than services. In addition, manufacturing generates far more value-added per worker and thus acts as an engine of economic growth.

In the USA, as we saw in chapter 6, a vigorous debate developed in the early 1980s between the proponents and opponents of a national industrial policy. Industrial policy advocates such as Harvard's Robert Reich argued for the creation of new government institutions, such as loan agencies, which would help to regenerate US industry. But critics such as members of the Reagan administration argued that most of the industrial policy schemes proposed focused too much on trying to "save" jobs in Rust Bowl industries rather than creating viable new ones. This amounted to little more than a "prop-up-the-losers" approach, which MIT economist Lester Thurow had dubbed "lemon socialism."

This is not to say that both sides in the industrial policy debate did not agree that manufacturing was still important: rather, the dispute was about how best its future might be secured. The opponents of government intervention argued that smokestack industries need not die completely, but they would have to become leaner, more productive and better managed. This would be best achieved by relying on the traditional American skills of entrepreneurship. The old industries would also be helped back to recovery by large doses of high-tech.

In the UK, the British government launched a campaign to boost technology in traditional industries, in recognition of the fact that the choice posed between "sunrise" industries on the one hand and "sunset" industries on the other was in reality a false one. The real task, argued government ministers, was to make traditional manufacturing competitive once again through the application of new technologies, as well as better design, marketing and new working methods. The need to revive British manufacturing became even more urgent as North Sea oil production peaked and it was learned, for instance, that the famed City of London earned only £1 for every £8 worth of exports by British industry, despite manufacturing's recent rapid decline.

Yet in the USA and Europe it was also becoming obvious that any expansion of domestic manufacturing – through a general "upturn" in the economy or through greater competitiveness – was unlikely to create many new jobs. The weight of evidence from analysts and from industry itself pointed to unemployment remaining high even if business picked up. Pretty soon the commentators came up with a handy phrase to describe the new phenomenon: we were into an era of "jobless growth."

This paradox came about because improved productivity and the installation of new technology reduced the demand for labor even as output increased. As a 1983 *Business Week* survey showed, there was also much spare capacity in manufacturing industry, and companies were not about to take on lots of new labor when they had only recently – and expensively – got rid of so much. The "miracle" of manufacturing productivity had a less attractive side to it.

A comprehensive study of the employment effects of new technology by Dr Dan Finn, published by the London-based Unemployment Unit in late 1984, argued that any increase in demand for goods was likely to stimulate the development of new production processes using new technologies that require fewer workers: "For these reasons, we are unlikely to see any significant growth in manufacturing employment." Professor John Constable, director-general of the British Institute of Management, was more blunt. He stated in June 1985: "I do not believe manufacturing is going to create any new jobs in this country for a very long time – or possibly ever again."

The Finn study concluded that, if economic growth coupled with reductions in working time could contain and reduce the unemployment problem, then the "optimists" will have been proved right. But the balance of the evidence so far indicated that the new technologies were reducing the demand for labor at a rate with which economic growth and marginal changes in working time could not cope. This suggests that the solution to unemployment does not lie with a general reflation: policies that had worked in the 1950s and 1960s are inadequate in the new era of jobless growth, and more radical solutions will be required.

What Kind of Work?

For millions of people, the world of work is changing under the impact of high technology and new working arrangements. Of course, changes in the pattern of employment are nothing new, but the cumulative effect in the long term can be dramatic. Back in 1830, 70 percent of the US workforce was employed in agriculture; by 1900 only 40 percent worked on the land, and today the figure is more like 3 percent. Likewise, the proportion employed in manufacturing has plummeted from over 50 percent to less than 25 percent in the last 50 years and could go as low as 10 percent by the end of the century.

With the spread of high technology, we are seeing a further restructuring of work which will have an equally dramatic effect on society – perhaps over a shorter time-scale. There has been no instant industrial revolution, no mass shut-out caused by automation alone; but the *pattern* of employment in Europe and America is changing swiftly and significantly. In fact, the occupational structure has become a shifting mosaic of industrial winners and losers, and the job market a mixture of dead-ends and bright prospects.

Top gainers in the USA between now and 1995 in percentage terms will be computer programmers and systems analysts, electronics engineers, accountants, lawyers and physicians. But the numbers involved are small compared with those in low-tech jobs, and these occupations do not even figure in the top ten growth occupations in terms of total numbers, which includes secretaries, clerks, nurses, janitors, cashiers and waiters (see figure 9.1). Top losers in the USA in percentage terms will be postal clerks (down 18 percent – victims of technology) and college lecturers (down 15 percent – victims of falling rolls and government policy). But the largest absolute declines will be recorded by farm operators and laborers, and by teachers.

Two thirds of the jobs created in the USA in the past decade have been taken by women. The proportion of adult women who are working

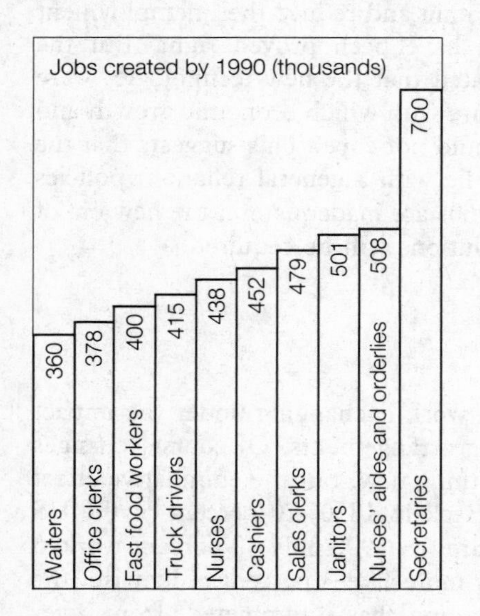

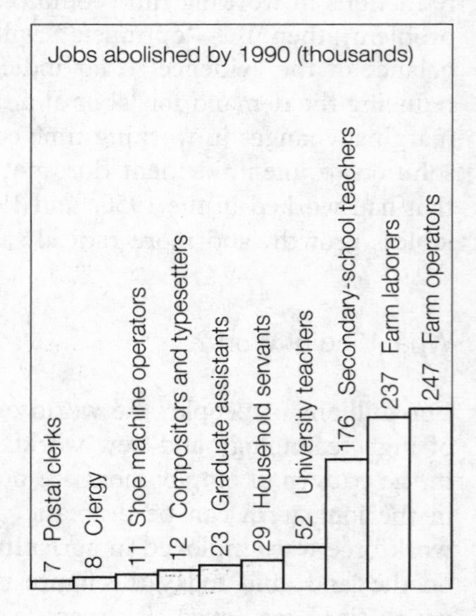

Figure 9.1 Where the jobs will – and won't – be.
(Source: *Fortune* magazine, May 16, 1983, p. 111. Reproduced by kind permission of *Fortune* magazine, © 1983 Time Inc. All rights reserved.)

(the female "participation rate") has grown to 54 percent (from 43 percent in 1970) and is expected to reach 60 percent by 1990. Women overwhelmingly choose the service industries, where they hold 80 percent of clerical jobs and 60 percent of jobs in the food and health sectors. But women's wages, at 64 percent of the male average, still fall far short of equality, and women also tend to be concentrated in occupations and industries that will be affected most by new technology. However, women have fared better in recent recessions and manufacturing slumps, which have hit men hardest.

Europe is broadly following the US trend. In the UK, for instance, the proportion of the workforce employed in manufacturing is down to around 25 percent, and a similar pattern of occupational winners and losers is apparent. In 1983 white-collar workers outnumbered blue-collar workers in Britain for the first time, just as the UK was becoming a net importer of manufactured goods for the first time since the Industrial Revolution.

But these are not the only changes taking place in the British way of work. As in the USA, there has been a big increase in self-employment, part-time working, temporary working, contract working, flexible working and homeworking: more of the British workforce is self-employed than at any time since 1921. In 1984 the total reached 2.5 million, a 32 percent increase over 1979. The self-employed now account for roughly one in ten of the UK workforce. As in the USA, part-time working has also grown – to one in five of the workforce – and the figure is expected to reach one in four by 1990. The total number of part-timers has doubled in the past 20 years. Two-thirds of them are women.

Flexible working arrangements are growing in popularity. This is creating new patterns of working time – shorter working weeks, flexible working hours, even flexible working years. There has also been a big increase in homeworking or "telecommuting" by professionals such as computer programmers. In Britain there are 1.7 million homeworkers, who now account for 7 percent of the workforce; in the USA, where the figure is more than 10 percent, some 240 companies are running homework schemes (as we saw in chapter 5).

Just as fears in the USA about the quality of many new service sector jobs have led to allegations that these (mostly part-time) workers are being treated as second-class citizens, so in Britain recent research has identified a growing gulf between what are called "core" and "periphery" workers. Inner "core" workers are full-timers with secure jobs, who enjoy good terms and conditions of employment. The outer "periphery" workers are typically on temporary contracts or are working part-time, are self-employed or are employed by subcontractors. Core workers are "functionally flexible": in return for security and decent conditions, they are expected to do whatever work the company demands. The periphery, in contrast, is hired to do specific jobs and is fired when not needed – they are "numerically flexible." This two-tier job system is set to grow in the future.

Despite this, suggestions in the USA that deindustrialization and the shift to services is demolishing the middle class seem to have been overblown, according to the latest evidence. In the early 1980s, some economists such as Barry Bluestone of Boston College were arguing that the loss of high-paying industrial jobs was helping polarize the occupational structure between a large number of low-paid workers and relatively few high-paid workers. As the AFL–CIO put it in its pamphlet, *Deindustrialization and the Two-Tier Society*, "The once solid middle tier of American jobs has been undermined."

But a closer look at what was happening in the mid-1980s suggests that the "vanishing middle class" was a myth, partly the result of a Census Bureau statistical fluke and partly due to demographic trends. Such evidence as there was to support the hypothesis could not be laid at the door of the service industries. The growth in white-collar and high-tech service jobs that pay average or above-average wages was fast enough to offset the loss of skilled blue-collar jobs. Moreover, the computer boom itself will play a major part in the enhancement of service jobs, both directly and indirectly, through new capital investment in the service sector.

While economists may continue to disagree about the "vanishing middle class," there is a broad consensus that pervasive information technology is changing the nature of work for almost every employee. Few jobs will remain unaffected by computers. Computer-mediated work demands new skills in dealing with abstract information systems, and there is less reliance on direct experience. "It isn't just the speed and efficiency of computers that will revolutionize work," says the Harvard Business School's Shoshana Zuboff, "the computer fundamentally changes the relationship between a worker and his task." At the same time, high-tech systems allow workers the flexibility to tackle different kinds of job and frees them to spend more time on creative work and problem-solving, rather than on performing dull, repetitive tasks.

One point all the experts agree on is that vast numbers of workers will need to be retrained in high-tech skills. Those without them will face a bleak future on the labor market. A British study on the social impact of information technology for the UK's National Economic Development Office (*IT Futures*, NEDO, London, 1985) concluded that the demand for higher-level information management skills will increase, while many lower-level skills (and occupations) will be incorporated into the emerging technologies. The demand for multi-skilled personnel will also increase.

Thus, according to the NEDO report, "Rather than leading to a workforce polarized into a highly skilled sector and an unskilled sector, there is an emerging consensus that the workforce as a whole will need to be more highly skilled." Technicians and maintenance workers will require software skills, while many semi-skilled and unskilled jobs will disappear (as we saw in the case of offices in chapter 7). In order to remain employable, craftsmen and draftsmen will need to be familiar with computers and electronics. All this will necessitate major changes in the UK education and training system, concludes the NEDO report.

"Retraining" has also come to be an issue in the USA. Indeed, it has become something of a rallying cry on Capitol Hill, where legislators have been caught up in the hysteria over foreign competition. Even conservative Congressmen have been calling for government intervention and higher spending on training. In the autumn of 1985, more money was voted to retrain laid-off factory workers in new high-tech skills.

As the future of work in the form of paid employment has become more problematical, some economists and sociologists have toyed with new definitions of work, asking the very basic question: What is "work" anyway? In particular, they have taken a new interest in what has been termed the "informal economy." The informal economy covers work done which is not traditional paid employment in the formal economy. The term was first used in a Third World context, where it was obvious that a great deal of unofficial economic activity was taking place "off the books" in agricultural produce markets, back street sweatshops and so on. The term was then applied to advanced Western societies, where it became apparent that unofficial economic activity had increased as the postwar era of full employment and mass consumption came to an end.

The informal economy consists of two main areas: the "grey" economy of work done for free at home – mainly housework, but also including do-it-yourself (DIY) tasks and the use of home, kitchen or garden gadgetry to produce goods or services for home consumption; and the "black" economy or hidden economy, consisting of work that should be part of the formal economy but is done for cash on the side or "off the books" (that is, without the Internal Revenue Service (IRS) knowing about it). As Professor Joseph Huber recently pointed out, there is a great variety of terms and concepts used to describe different aspects of the informal economy; suggestions have included the "green economy" to cover voluntary work in the community, and even the "mauve economy" of personal services such as "kiss-o-gram" greetings.

The black economy is small and illegal, while the grey economy of the household and the community is legal and large, involving all adults. Professor Peter M. Gutman of the City University of New York estimates that the black economy in the USA amounts to 10 percent of GNP, while the IRS (which has an obvious interest) says it is worth between 6 and 8 percent of GNP. The IRS also calculates that about 4.5 million people in the USA live entirely on their earnings from unofficial jobs.

In the case of Britain, Professor Richard Rose of Strathclyde University has estimated that the grey economy might account for 51 percent of all labor hours, compared with 46 percent in the formal economy and only 3 percent in the black economy. Other estimates have put the size of the black economy as high as 7.5 percent of GNP in Britain, while in Italy it is thought to be worth as much as 20 percent of GNP. But many now think that in most of Europe the death of the formal economy has been greatly exaggerated.

There are broadly two opposing views of the black economy. One is that it offers an alternative form of work which has the potential to absorb unemployment by encouraging entrepreneurial initiative, and is therefore a good thing; the other, more critical, view sees it as less significant and if anything as a sign of desperation, a kind of survival mechanism for the poor, rather than as something pointing the way to an alternative society.

Thus, some have argued that the black and grey economies are bound to grow as the unemployed and the underemployed do more work on the side for others and more DIY-type work for themselves around the house. But this theory has been knocked on the head by R. E. Pahl's recent study of households on the Isle of Sheppey in Kent, England (*Divisions of Labour*, Basil Blackwell, Oxford, 1984). Pahl found that it was the *already employed* who did more work around the house and carried out more "self-provisioning" such as wine-making and beer-brewing: the unemployed were apathetic, short of cash and lacked the necessary DIY equipment.

Not only is the grey or household economy much bigger than the black economy, but it may have a much longer-term significance for the future of paid employment. For example, in *After Industrial Society? The Emerging Self-Service Economy* (Macmillan, London, 1978), Jonathan Gershuny pointed out that capital goods in the home such as washing machines, dishwashers, freezers, food processors, microwaves, paint-strippers, hover-mowers and VCRs were being substituted for services purchased outside the home from laundries, restaurants, decorators, gardeners and cinemas. Any increase in such "self-servicing" would clearly have implications for the future of service sector employment.

Extending the argument in *The New Service Economy* (Frances Pinter, London, 1983), Gershuny and Ian Miles argued that the new information technology, especially the new telecommunications infrastructure of digital telephones and cable, will bring about a further wave of social innovation, for example by producing new services such as business information services run from the home. More people working from home could have a mixed impact on jobs: it could reduce the demand for some services in the formal economy, such as transport, while generating new jobs in the informal economy, which will in turn generate demand for products of the formal economy such as word processors, printers and so on. The UK NEDO report, *IT Futures*, devotes a section to exploring these links between the household economy, consumption and economic growth.

Whatever the truth about the exact size and significance of the grey and black economies, we should remember that there is more to work than paid employment. Jobs are important, but they are not the only way a person can contribute to society, make friends or find an identity.

Managing Technological Change: Will Labor Be Left Out in the Cold?

The high-tech revolution presents a major new challenge for managers. The arrival of information technology (IT) is changing the nature of a manager's job and is spurring the adoption of new management styles and practices which have the potential to transform conventional adversarial management–labor relations. IT is also bringing about the internal restructuring of companies, by flattening the management pyramid, and the restructuring of whole industries, by altering their boundaries and the rules of competition. As for the unions, IT presents them with perhaps their greatest challenge yet: afflicted by outdated attitudes, paralyzed by an ambivalence toward technological change and beset by adverse social trends, labor in the high-tech society is in severe danger of being left out in the cold.

The advent of desktop computers has put masses of information at the elbow of managers. Spreadsheets, databases, electronic mail, voice-store-and-forward telephone systems and other gadgetry are boosting management productivity. As speech synthesis and voice recognition technologies become available, managers will be able to "talk" to computers and get instant voice responses. As product life-cycles get shorter, as the pace of innovation hots up and as the need to move quickly into new markets becomes paramount (viz., the computer industry itself), there will be an increasing need for high-speed management to be equipped with these tools of the high-tech age.

One ever-present danger is that of information overload or "info-glut": managers can be deluged with so much finger-tip information that they can't see the wood for the trees. As Paul Strassman argues in his book, *Information Pay-Off* (Collier Macmillan, New York, 1985), throwing technology at a problem is not always the answer. After studying several groups of companies, he found no correlation between spending on IT and management effectiveness. Instead of making individual managers more effective, the heavy use of computers in some firms had proved a disaster and had led to the employment of extra staff to handle all the data. But careful planning can avoid this, says Strassman.

A common experience of companies is that information technology eliminates some of the layers in the corporate structure. The management pyramid becomes progressively flattened as those at the upper levels gain instant access to information which previously had to be extracted laboriously by those at lower levels. Moreover, it is no longer necessary to rely on vertical, hierarchical relationships: smaller, more cohesive and cooperative management structures are made possible by IT. It is also possible to create better communications and greater integration between departments – provided that the correct design choices are made.

Apart from the internal restructuring of companies, IT can bring about the restructuring of whole industries by changing product costs and values and by making it easier (and sometimes more difficult) for newcomers to enter the business. As Michael Porter and Victor Millar of the Harvard Business School have shown, IT spawns completely new businesses and, perhaps most important of all, provides a lever that companies can use to gain competitive advantage over their rivals. Thus, information technology is fast becoming an indispensable strategic tool in marketing, customer service, training, product development and strategic planning. It has already had a tremendous impact on the nature of the banking, distribution, travel and airline businesses, revealing entirely new approaches to existing markets and creating whole new product lines.

The introduction of IT is changing the way managers handle the workforce. The old "kick ass" routine is no longer applicable to today's increasingly knowledgeable and highly skilled workforce, who have greater autonomy and spend more time solving problems than performing routine tasks. IT has stimulated a great rush to adopt more

enlightened, participatory management techniques – many of them, such as "quality circles," pioneered in Japan, Scandinavia and West Germany. These alternative management practices are designed to reduce or end the "them"-and-"us" confrontation and to pave the way to a more appropriate method of conducting industrial relations in the high-tech age.

But first, the actual installation of IT systems has to be handled with great care. As Boston University's Fred Foulkes and Jeffrey Hirsch have argued (in Tom Forester (ed.), *The Information Technology Revolution*, Basil Blackwell, Oxford, 1985), managers who wish to introduce robots successfully must carefully select their sites, move slowly, retrain displaced workers and educate and keep informed both line-managers and labor. Harvard's Shoshana Zuboff has warned that, if IT is to live up to its promise, managers must involve their workforces and consider the consequences of its arrival for human beings and the quality of their work environments.

One point made frequently in the literature is that a particular type of technology does not determine that a particular form of work organization will be adopted or a certain quality of worklife achieved. Much depends on how and why IT is used, which in turn depends on the prevailing economic and social conditions and the dominant managerial ideology.

Despite the impression sometimes given in the popular media, most studies report very little effective trade union resistance to new technology. On the contrary, management resistance seems to have been more of a problem – that was the conclusion, for example, of a survey for the UK National Economic Development Office edited by Peter J. Senker (*Planning for Microelectronics in the Workplace*, Gower, Aldershot, England, 1985).

In another study, entitled *Unions, Unemployment and Innovation* (Basil Blackwell, Oxford, 1986), Eric Batstone and Stephen Gourlay of Nuffield College, Oxford, found that some changes had been abandoned because of union opposition, principally in telecommunications, printing and the civil service. But market factors – chiefly the recession – were a more potent reason for non-implementation. There is also some evidence to suggest that disputes over new technology are more common in the public sector. Most unions in the private sector in Europe seem to have accepted the "automate-or-liquidate" rationale behind the adoption of IT.

In fact, labor unions in Europe have taken such a beating in the old smokestack industries that there is growing evidence of a major change in attitudes – characterized by a greater willingness to accept "no strike" agreements (especially with Japanese firms) and "single union" deals, and to sign contracts that provide for flexible working. "Single union" deals – which mean that all categories of worker in a company or plant belong to the same union – clearly have implications for union structures.

Faced with a huge loss of members in Rust Bowl areas, some UK unions such as the electricians (EETPU) have launched recruiting drives in the high-tech industries around Cambridge, along the "M4 corridor" or Thames Valley area, in Scotland's "Silicon Glen" and in the newly revolutionized Fleet Street, home of Britain's newspaper industry. These efforts have met with mixed success – most computer companies remain staunchly non-union – and there is evidence to suggest that high-tech firms are purposely avoiding the older industrial areas where the unions are strong. The recruitment drives have also provoked rows within the British trade union movement, with critics charging that unions like the EEPTU, GMBU and ASTMS have so emasculated themselves in order to gain entrance to the "sunrise" industries that they are no better than the staff associations they seek to replace.

In the USA, organized labor is in even greater danger of getting left out in the cold. US unions have had little impact on the forward march of technological change in the older industries and they are poorly represented in the high-tech industries. As an example of labor's weakness, when General Electric set up its "factory of the future" in Lynn, Massachusetts, GE's management was able to impose just about everything it wanted on the International Union of Electrical Workers (IUE) in terms of new shift rotas, new job classifications allowing for flexible working and new production goals – all under the company's threat of withdrawal from the area.

In some industries, such as telecommunications, air transport and buses, deregulation and/or import penetration has added a further twist of the screw on the unions. When AT&T was divested of seven regional companies, this broke up the cozy old world in which higher labor costs were regularly passed on to consumers. And the Communication Workers of America (CWA) just grew as AT&T did. But the new regional companies are not nearly so pro-union, and the cold winds of competition are keeping down wage increases.

A recent study of the automobile industry, by Harry C. Katz of Cornell University and Charles F. Sabel of MIT, suggests that the new flexible manufacturing systems are undermining traditional management–labor relations. National unions can no longer guarantee industry-wide job descriptions and work rules. In the new high-tech era, unions instead will participate in making basic business decisions, will promote employment security and will give their members access to the diversity of skills they will need. This approach is similar to the labor relations practices of the more successful Japanese and West German car makers, who have shown themselves to be far more advanced in the management of technological change.

High-Tech Crime: Are Hackers Heroes?

Computers have created opportunities for crime that never existed before. The rapid spread of personal computer terminals and distributed processing have made high-tech rip-offs much easier for all grades of white-collar worker. The new technology is thus *democratizing* white-collar crime, because it enables even the humblest programmer or operator to participate in illegal activities that were once pretty much the preserve of top management.

The actual losses incurred through computer crime are difficult to ascertain. In the UK, estimates vary from £500 million to £2 billion a year. A 1986 government-sponsored study found that 40 percent of large UK companies had suffered at least one major computer fraud in the previous ten years. Reported annual losses in the USA were running at the level of $100–$300 million in the late 1970s, but more recent estimates – which include both reported fraud and a figure for unreported fraud – have put annual US losses as high as $3–5 billion.

The American Bar Association (ABA), in a major report published in 1984, concluded that losses arising from computer crime sustained by US business and government institutions were "by any measure, huge." An analysis of 300 of America's top corporations suggested that annual average losses per company could range from $2 million to as high as $10 million. "If the annual losses attributed to computer crime sustained by the relatively small survey group are conservatively estimated in the range of a half a billion dollars, then it takes little imagination to realize the magnitude of the annual losses sustained on a nationwide basis," said the report.

Many experts believe that only about 10 percent of computer frauds are made public by companies and that many crimes go completely undetected. Those that are detected are rarely prosecuted: a 1986 National Center for Computer Crime Data survey found that fewer than 100 cases of computer fraud had been prosecuted in the previous two years. What is clear is that nobody is very clear about the exact size of the problem – but most agree that it is large and growing. Even if the percentage of installations affected may be small, the sum involved in the average computer crime is much larger than in conventional robberies. As to the future, the ABA report concluded: "It would seem beyond dispute that computer crime is today a large and significant problem with enormous potential for becoming even larger and more significant."

Computer crime has been defined as a criminal act that has been committed using a computer as the principal tool. The rapid growth in the number of people with computer skills means that there are many more potential computer criminals. Computer crime can include the theft of money (for example, by making transfer payments to the wrong accounts), the theft of information (by tapping into data transmission lines or into databases at no cost), the unauthorized use of computers for private purposes, and just plain sabotage or game-playing for the hell of it – the (mainly young) people who do this are called "hackers."

Banks are a major target for the new breed of high-tech criminals. Like everyone else, banks are vulnerable to crimes committed by employees and to crimes committed by outsiders playing "vault-invaders." The increased reliance on electronic funds transfer (EFT) systems – $200 billion changes hands daily in the New York banks' automated payments system – has increased the risk of fraud on a massive scale. If the electronic authorization codes used in EFT fall into the wrong hands, huge sums of money can be moved about – including out of the country – in a matter of seconds.

In a famous case in 1979, for example, Stanley Mark Rifkin, a computer consultant to the Security Pacific National Bank, visited the bank's wire transfer room where he learned the EFT codes. Later, posing as a branch manager, he phoned the Los Angeles bank and used the codes to transfer money, in amounts of less than $1 million, to a New York bank. Then he instructed the New York bank to send the money – by now totalling $10.2 million – across to a Swiss bank account. Having flown to Switzerland, he converted the money into diamonds and then returned to the USA. It was only when he boasted openly of his feat that he was caught and convicted.

Other kinds of computerized bank fraud include fictitious loans, unauthorized lines of credit and various other forms of transaction manipulation. The American Bankers' Association is wary of putting a figure on computer rip-offs from banks and says that such frauds should be seen in the context of the $2 billion dollars lost in 1985 through credit card fraud alone. In the case of Britain, reliable information on bank losses through computer fraud is almost impossible to obtain, but it seems that the major banks set aside about £85 million to cover them, and it is believed that both the Midland and the National Westminster Banks have recently suffered multi-million-pound heists. However, CHAPS (the Clearing Houses Automated Payments System) is thought to be one of the most secure financial systems in the world.

In 1985 a computer fraud survey by the Audit Commission for Local Authorities in England and Wales concluded that high-tech crime could be costing the UK millions – but there was no way of really knowing, because most frauds go completely unreported. The Commission stressed that banks and other financial institutions were particularly keen *not* to admit to having been conned and preferred to cover up crimes, fearing loss of customer confidence.

Computer thieves have also been attracted to insurance companies. One of the largest computer crimes so far discovered involved an insurance company, Equity Funding, which used its computers to generate thousands of phony insurance policies that were later sold to re-insurance companies for a total of over $27 million. In the UK, fraud is costing insurance companies £2 million a day. According to an American Institute of Certified Public Accountants (AICPA) study, other kinds of insurance computer crime include fictitious claims, fraudulent loans against policies and the cancelling of policies to gain premium refunds. Computers are also being used to steal goods by altering inventories and re-directing items, or to steal information contained in company records and databases. The theft of copyright software is almost routine.

The methods used to burgle a computer include taking advantage of an error in a program and making deliberate changes to a program. This places great temptation in the way of programmers and systems programmers, who are often the only ones who know how a system works. With control of a computer's operating system, a programmer can copy or change data and programs, and restore them after the completion of their deed.

In fact, computer criminals use a variety of techniques which have brought forth a whole new jargon of software sabotage:

- *The Salami* This technique involves spreading the haul over a large number of small transactions, like slices of salami. For instance, a bank clerk might shave a trivial sum off many customer accounts to make up a large sum in his or her own account.
- *The Trojan Horse* This involves the insertion of false information into a program in order to profit from the outcome – such as a false instruction to make payments to a bogus company.
- *The Time Bomb* and the *Logic Bomb* These terms refer to the insertion of routines which can be triggered later by the computer's clock or a combination of events. When the "bomb" goes off, the entire computer system – perhaps worth millions – will crash and be rendered useless. These two techniques can be used as the basis for ransom demands.
- *Data Diddling* This involves altering the data being fed into the computer or swapping one piece of data for another in order to benefit the fraudster.
- *Zapping* and *Superzapping* These terms refer to the penetration of a computer by unlocking the master key to its program and then destroying it by activating its own emergency program.
- *Piggy Backing* This entails tapping into communications lines and riding into a system behind a legitimate user who has obtained password clearance.
- *Scavenging* This process involves searching through stray data or "garbage" for clues that might unlock the secrets of a system.
- *Worms* Similar to zapping, worm programs delete specific portions of a computer's memory, thus creating a hole of missing information, or else alter a system's operations or shut it down completely.
- *The Virus* This is a program that instructs the host machine to summon up its stored files. It then mixes them up, turning the computer's memory into a mass of confusion.

- *The Trapdoor* This is a method of collecting legitimate users' passwords as they log on. No matter how often the users change their passwords, the trapdoor intervenes and reads off the new list of passwords, thus allowing continued unauthorized access to the computer.

"Hacking" means using a personal computer and modem to enter other people's computer systems over the telephone wire by finding the right password. These "computer break-ins" are usually carried out by young computer enthusiasts with a sense of fun, who see it as a challenge to crack the secret codes of computer installations. Famous examples include the kids of Manhattan's Dalton High School, who in 1980 used their classroom terminals to enter a Canadian data communications network, destroying key corporate customer files in the process. In 1984 young West German hackers caused chaos when they broke into their country's videotex network, and some French hackers actually gained access to the secret files of France's nuclear programme.

Hackers – sometimes also called "modem maniacs" – have become increasingly controversial in recent years. Some see them as modern folk heroes, while others regard them as little better than common criminals. The publication of two how-to-do-it books, Bill Landreth's *Out of the Inner Circle* (Microsoft Press, Bellvue, Washington, 1984) in the USA and Hugo Cornwall's *The Hacker's Handbook* (Century, London, 1985 – revised as *The New Hacker's Handbook* in 1986) in the UK have added fuel to the controversy.

Landreth is a young hacker who sees getting to the top of complex computer systems as a great challenge. In his book, he explains how simple it is to find by trial and error the obvious and easy-to-remember passwords that companies and government organizations often use in their systems. In doing so, he reveals an astonishing lack of security: "Computer owners and system operators," he says, "should take the time to educate their users. If they did, hacking as it is today would fall to such a low level of activity it could be considered dead. It's that simple."

Hugo Cornwall, whose book was condemned by Scotland Yard, the HQ of London's Metropolitan Police (thus ensuring it became a bestseller), takes a similar line in warning of the astonishingly lax procedures adopted by systems operators, after first explaining how hacking is done. He argues that hackers are merely the latest in a long line of tinkerers who like playing around with machines and technology. Steven Levy, in *Hackers: Heroes of the Computer Revolution* (Doubleday, New York, 1985), goes even further in his celebration of the breed, as the title of his book implies.

Cornwall says the pleasures and rewards of hacking are all intellectual, and he urges hackers not to cause damage to systems or commit fraud. He also points out that many crimes classified as "computer fraud" have little to do with computers as such and are mostly committed by insiders rather than outsiders. But Cornwall, like Landreth, fails to show clearly where high-tech high-jinks end and high-tech crime begins.

Combating computer crime calls for new kinds of security measures – measures that often include the use of computers. Many companies are extremely lax in their computer security, often believing that computer frauds could never happen to them. The problem with computer security is that everyone talks about it but not enough people do something about it. Bank security and industrial security are well understood – computer security is not. Most experts lay the blame for poor computer security squarely at the door of top management. Often they can't be bothered, or they don't wish to restrict ease of use, or they don't appreciate what their information is worth.

Computer security can be greatly improved simply by using a bit of common sense. The passwords allowing access to systems, for instance, can be made less obvious and memorable by avoiding such passwords as girlfriend's names. Passwords should be issued only to the absolute minimum number of people requiring access.

A growing market is now developing for access-control software that closes password loopholes. This software restricts users – individually identified by passwords and codes – to only those files they are authorized to use. Even then, the software permits users to perform only authorized functions, such as adding or deleting information. One major limitation with access-control software, however, is that it does not protect a company against frauds committed by employees while going about their legitimate tasks.

Also growing more popular are "black box" techniques or "dial-back" systems. When a user calls into a computer, a black box intercepts the call and asks for a password. The unit then disconnects the call, looks up the password in the directory and calls the user back at his or her listed telephone number: fraudsters calling from another number will be screened out.

Scrambling devices and encryption software are being developed which scramble messages for transmission so that only the legitimate recipient can understand them. Anyone tapping into, say, a bank's communications line will find a scrambled list of zeros and ones. However, even the best encryption codes or "keys" can be broken – a lot depends on the length of the key. In 1986, the National Security Agency announced that it was changing its coding system for data used by government agencies.

The Pentagon not only changes its codes daily – and even hourly, for the most sensitive information – but also encases its private telephone lines in metal tubes filled with high-pressure gas. A sudden loss of pressure will signify a break in the tube caused by an unauthorized tap. Under its "Tempest" program, the Pentagon is spending $200 million a year to eliminate or muffle signals from machines used by the military, security agencies and defense contractors.

Audit software packages are now available which can monitor transactions or the use of a computer. These enable auditors to trace and identify any operator who gains access to the system and when this occurred, such as after-hours. Audits can also highlight an abnormal number of correction entries, which often indicates fraudulent activity. The demand for auditors with computer skills is high, and there are not enough who are capable of outsmarting crooked computer personnel.

Computers are also being used increasingly in the fight against conventional crime. One of the most exciting areas is the use of computers for fingerprint identification using digital techniques. One such system, developed by NEC of Japan, cracked 34 unsolved cases in its first few days of operation with the San Francisco Police Department. After the first ten weeks it had solved 360 crimes, including the 15 murders committed by the so-called "Night Stalker." A suspect print lifted from an automobile was compared with 380,000 others stored in the computer's memory. It then came up with ten names: at the top of the list, with a probability rating four times as high as the nearest contender, turned out to be 25-year-old drifter, Richard Ramirez, who was subsequently caught and charged with the murders.

This new science of digitizing biological characteristics has been called "biometrics." It embraces not only computerized fingerprinting, but also hand geometry, voice recognition, vein scanning and retinal scanning. Devices that scan fingerprints can be used to control access to military bases, bank vaults and stores. They are much less open to abuse than door keys, digital locks and card systems. An Oregon company, EyeDentify Inc., has come up with a "What the Butler Saw" device which scans the eyes and analyses digitally the pattern of the blood vessels in the retina. This is supposed to be even more accurate than scanning fingerprints, which can be smudged or can suffer from cuts and abrasions.

The police are using computers in all kinds of new ways – from information storage to electronic navigation systems in patrol cars. In the war against counterfeiters, original manufacturers are using high-tech techniques such as hidden microchip tags, disappearing–reappearing inks, holographic images and digitized labels that are hard to copy. The legal system is being transformed by the use of computers in administration, court reporting and the creation of legal databases containing records of every important court case.

"In fifteen years' time," said a Scotland Yard man recently, "almost every crime will involve a computer." He could have added that almost every crime solved will be *solved* with the aid of a computer. Even so, unless computer security is taken much more seriously – and the theft of information made illegal – computer crime and software sabotage could cause chaos in the 1900s.

The Threat to Privacy

The new information technology makes it possible to store, retrieve and analyze masses of "transactional information" – records of phone calls, credit card payments, air travel and so on – in huge databanks. Personal data collected by private corporations, government agencies and other organizations can now be stored on a vast scale, thanks to cheap computing power. What's more, information stored in one automated dossier can be correlated with information in other databases and can be transmitted around the country in seconds at relatively low cost. The threat to privacy could not be more obvious.

The computers of the five largest credit-checking companies in the USA, for instance, contain records on more than 150 million individuals. The banks have records of the hundreds of millions of checks written and electronic fund transfers made each year. Telephone companies can provide computer printouts of calls made and calls received. Airlines and car rental companies operate computerized reservation systems that effectively keep track of the movements of millions of people. The health records of nine out of ten working Americans are computerized through their insurance policies, while even drugstores are now keeping records of the drugs prescribed to each patient.

According to David Burnham, in his book *The Rise of the Computer State* (Random House, New York, 1983), the US government has collected 4 billion separate records about the American people – 17 items for every man, woman and child in the country. For example, the tax records of 100 million Americans are held on Internal Revenue Service (IRS) computers. Federal agencies, not counting the Pentagon, have at least three separate telecommunications networks covering the whole of the USA. The FBI, through its computerized database, the National Criminal Justice Information Center, has files on the millions of Americans arrested each year even for the most trivial offences. State and federal governments have established more than 500 programs to compare information from two or more sources.

The value of transactional information was dramatically illustrated a few years ago when data from the pre-deregulation AT&T provided a complete record of the telephone calls between President Carter's brother Billy and the government of Libya. It was enough to sink a very special relationship – and contributed to Carter's loss of the presidency. Then there's the matter of the missing telephone records of New England Bell, which would have revealed who Teddy Kennedy called immediately after the accident at Chappaquiddick Creek that claimed the life of Mary Jo Kopechne. . . .

The National Security Agency (NSA) can simultaneously monitor 54,000 telephone calls made to and from the USA. The Agency operates outside of judicial or legislative controls. The Crime Control Act of 1968 makes it a felony for a third party (except such government agencies) to place electronic listening devices on telephones. But this law does not apply to *information* transmitted in digital form – and an increasing proportion of calls, especially long-distance calls, are transmitted in this way. Furthermore, US privacy laws do not cover cellular mobile phones, electronic mail, databanks and videoconferencing – although two bills before Congress in mid-1986 hoped to remedy this.

The rise of digital transmission has also created a national security problem because it has made electronic eavesdropping by the Soviets much easier. In 1984 President Reagan directed the NSA to investigate ways of creating more secure telephone networks for the communication of sensitive government information. But the cost of installing "secure" phones with the help of scrambling devices is high. Meanwhile, in the UK British Telecom (BT) in 1986 unveiled an encryption chip device that scrambles information into garbled junk before it is sent over telecommunications lines. BT is also working on another encryption chip which would make satellite data transmission more secure.

The *accuracy* of much of the information stored on databases such as those of the FBI has been frequently challenged. For example, in 1981 the Office of Technology Assessment (OTA) commissioned New York criminologist Dr K. C. Laudon to make a study of the value of criminal history data contained in the FBI and state police agency files. He found that a high proportion of the information was incomplete, inaccurate and ambiguous. A great deal of it involved arrests and investigations that did *not* result in a conviction or related to minor offences in the dim and distant past. Other studies have shown that employers are most unlikely to employ such people with a "criminal record." Four out of five states approached by the OTA admitted that they never checked the accuracy of the data in their files or conducted regular quality audits.

For decades, civil liberties organizations have campaigned for the right of individuals to see and correct computer records on themselves. Sweden led the way with data privacy legislation, while the USA followed with the Privacy Act of 1973 which established the principle of public access. The much stronger Privacy Act of 1984 requires that individuals be notified that personal records are being kept on them, gives them the right to see and correct them, and prohibits information provided from being used in an unrelated context. But this applies only to federal records, and it has yet to be extended to state, local and private sector records.

In the UK, where there are some 220 government databases alone, each containing between 10,000 and 1 million names – and where there are 5 million names on the Police National Computer – a Data Protection Act was finally passed in 1984. This requires all organizations who hold information on computers about individuals to register the type of data they hold with a Data Protection Registrar. The Act enables citizens to obtain compensation via the civil courts if the personal data held on them is found to be inaccurate, or if the record is lost or is disclosed to an unauthorized person.

But critics say that the 1984 Act is feeble compared with data protection legislation passed in other countries, and it fails to reflect the recommendations of the Younger Committee on Privacy (1972) or the Lindop Report on Data Protection (1978). For instance, manually operated files are not covered by the Act, and there is also a long list of exceptions, including data held by the police, by the Inland Revenue, by immigration control and for "national security" purposes. There are in addition restrictions on access to data held by government agencies, health authorities and social services departments. The administration of the Act has turned into a bureaucratic nightmare for the Data Protection Registrar.

In West Germany, the government actually passed a bill in 1986 legalizing computer-readable identity cards and passports. All West Germans will be required to carry the new ID cards, and this will permit the authorities to store data on the movement of people – including, say, peace campaigners – on central computers. A data protection law was enacted at the same time in the hope of ensuring that the new data would not be misused. But opponents, including data protection officers in Bonn and in the regional governments, continued to fight the plans, claiming the cards were ushering in a police state; similar claims were made in the USA when the FBI started tracking the movements of political activists in the early 1970s.

Even without ID cards and computerized passports, modern surveillance techniques can be used to monitor people's movements from aircraft, helicopters and even satellites. One-way video and film surveillance has spread rapidly – particularly in banks and shops – while hidden mikes, tiny tape recorders, miniature video cameras, sensors, voice analyzers, night vision devices and other chip-based surveillance gadgetry have become more widely available for use in commercial and international espionage. "Beepers" can be attached to offenders to track their movements, and sensors can be placed under the road to monitor the movements of specific vehicles. The last decade has seen increased use of polygraphs or "lie detectors" – which to a great extent rely on people *believing* that they work – and voice-stress analyzers, which allegedly can be used as hidden lie detectors.

MIT sociologist Gary T. Marx says that the new covert surveillance technologies conquer distance, darkness and physical barriers. They are capital-intensive rather than labor-intensive and can yield more information about more people at lower cost. Such information can be easily stored, retrieved, analyzed and communicated. Together with the growth of undercover "sting"-type operations and rewards for spying on colleagues or turning in suspected wrongdoers, Marx argues that the new "surveillance society" represents an unparalleled threat to privacy and thus human freedom. As Herbert Schiller argued, in *Who Knows: Information in the Age of the Fortune 500* (Ablex, Norwood, NJ, 1981), the new surveillance is lessening the power of individuals relative to large organizations and governments.

While the information-gathering powers of the state and private companies have grown with the spread of databanks, we have, Marx says, surrendered traditional notions of privacy for the sake of efficiency: "Inefficiency is losing its role as the unplanned protector of liberty." However, Marx does not argue that more harm than good now comes from these technologies. His point is that, in our drive to innovate and our infatuation with efficiency in day-to-day transactions, the potential for negative results has not received enough attention. More should be done to safeguard privacy now and in the emerging high-tech society. As Louis Brandeis put it, "The greatest dangers to liberty lurk in insidious encroachment by men of zeal, well-meaning but without understanding."

Problems to Come?

The future of work, management–labor relations, high-tech crime and data privacy have all been discussed in this chapter, but this is not to say that there won't be other problems facing high-tech society. There is, for instance, the major worry that high-tech society will become divided, as Kurt Vonnegut predicted in *Player Piano* (1953), into a highly skilled, highly paid technological elite of "haves" and an unskilled, unemployable underclass of "have-nots," with a growing gap between the information-"rich" and the information-"poor."

There is the problem of whether high-tech society will see information – and thus power – increasingly centralized (the pessimists' view) or decentralized (the optimists' view), with all the consequences for urban planning and society as a whole that probable changes – like an increase in homeworking – might entail. Politics will be affected by the new technology for good or ill – on the one hand by possibly increasing media domination and on the other by making possible a genuine participatory democracy.

The new information technology appears to be playing a de-massifying role in the structure of employment and in relation to organizations like labor unions and political parties. It should also make decision-making easier. But there is a danger that *too much* information, more cheaply provided, could make life more difficult by leading to "information overload" or a state of "info-glut." Waiting for all the facts to come in can be paralyzing when the facts never stop coming!

Finally, the high-tech revolution could aid technology transfer to the Third World, or it could exacerbate the already obscene inequalities between the rich North and the poor South. It will help us conquer the "High Frontier" in space – if we want to – and it will enable us to start (or prevent) a nuclear war much more easily. These are among the other key problems facing those who will take the decisions in high-tech society.

10 Conclusion

America at the Crossroads – What Will Japan Do Next? – Can Europe Catch Up?

America at the Crossroads

"If the United States continues on its present course, it will probably lose its status as the world's leading economic power before the end of the century." That was the startling conclusion of a recent editorial in *High Technology* magazine. Similar warnings have been delivered with increasing regularity by influential journals like *Business Week*, by leading academics in their books and articles and by worried politicians on Capitol Hill. All seem agreed that the USA now stands at the crossroads: either it can fight to maintain world economic leadership, or it can sink ignominiously into second-rate economic status – perhaps even within a few years.

The Japanese assault and the American retreat in world markets is now a familiar story. Japanese manufacturers have beaten US producers not only in consumer goods like radios, calculators, watches, typewriters, motorcycles, outboard motors, lawn mowers, cars, cameras, TVs, VCRs, CD players and camcorders, but also in heavy industries like steel, shipbuilding, machine tools, bearings and power shovels. The USA still does well in food, chemicals, aerospace, medical equipment and gas turbines. But US leadership in information technology is fading fast: the country is still ahead in specialized chips, automated data processing, supercomputers, software and telecommunications, but Japan already has world leads in standard chips, robots, fiber optics and office products like electronic typewriters and copiers.

Japanese success in industry has created a new awareness in the USA of the importance of manufacturing, which accounted for 30 percent of US GNP in 1953 but only 21 percent in 1985 (durable goods alone fell from 18 to 12 percent). It is now generally agreed that making things for sale in world markets is the key to national prosperity and that the USA must regain its competitive edge in manufactured goods by revitalizing its industrial base. The idea that the future lies in services, and that a post-industrial America can become more prosperous as a service-based economy, has been exposed as a dangerous myth. Manufacturing and services are largely interdependent, few services can be exported, and service jobs are generally low in pay and productivity. The USA still needs a strong industrial base.

The Presidential Commission on Industrial Competitiveness, chaired by John A. Young, president and chief executive officer of Hewlett–Packard, outlined the problem as follows when it reported in 1985. The USA ran a positive balance of trade from the turn of the century until 1972; then it plunged into deficit and the deficit grew steadily to $125 billion in 1984 (and to nearly $150 billion in 1985), with the US–Japan trade imbalance in electronics alone ($15 billion in 1984, $18 billion by 1985) becoming even larger than that in cars. US productivity growth had been either low or non-existent, with US gains being easily surpassed by Japan, South Korea, West Germany, Canada and even the UK; US real wages grew continuously between 1945 and 1973, but they had been falling steadily and now, average gross weekly earnings in real terms actually stood at 1962 levels. Finally, the US average real return on manufacturing investment had slumped from 12 percent per annum in the 1960s to about 4 percent in the mid-1980s.

Possible solutions suggested by Young and his team included boosting manufacturing investment by lowering the cost of capital. This could be achieved by reducing the federal deficit to cut interest rates, restructuring the tax system and creating a more stable monetary policy. Human resources could be better harnessed by putting more cash into vocational training and retraining. And international trade policy could be reformed under a single Cabinet-level Department of Trade which should re-examine such issues as high-tech export controls. But protectionism was seen as no real alternative to the production of desirable goods: as Harvard professor Michael E. Porter, a Commission member, commented, when a Japanese buying delegation scoured the USA looking for superior US goods to take home, it came up with just four items: wine, oranges, fondue sets and air fresheners!

Most interesting from our point of view, the Commission singled out technology as a key to international competitiveness and as the area where the USA still had its greatest strength. Pointing out that the percentage of US GNP spent on civilian R&D was lower than that of West Germany and Japan – and that the spillover from defense research was "incidental at best" – the Commission called for a complete reorganization of federal R&D under a new Department of Science and Technology. It also advocated tax credits to stimulate R&D, both to create new technologies and to apply existing technologies to manufacturing industry.

As Commission member Ian M. Ross, president of AT&T Bell Laboratories, argues, you can tinker with economic policy in this and that respect, but unless a nation stays at the cutting edge of high technology it is effectively finished as a potential world leader. High technology is all-pervasive, says Ross, and it has an elevating effect on other industries and services. Synergy between high-tech and traditional industries and services keeps the whole economy competitive: for instance, success in the automobile industry these days requires a thriving semiconductor industry, and vice versa. Although total US investment in R&D has grown in the 1980s to its highest level since the 1960s, more of it should be commercially orientated, and more of it should be directed toward manufacturing industry – yet future funding under the Reagan administration is now in doubt.

Faced with falling living standards and the phenomenon of "deindustrialization" – a declining proportion of the workforce employed in productive industry – the USA desperately needs to rebuild its manufacturing base. Apart from better management and a renewed desire to win, US manufacturing industries require massive doses of high technology if they are to survive and to overcome the new evil of "hollowing-out."

"Hollowing-out" occurs when companies shift production overseas to low-wage centers and/or they become mere assembly plants for foreign-made parts or re-labeling agencies for foreign-made products. Even the domestic automobile industry is not immune from hollowing: in 1986, 18 percent of parts in US-made cars were imported, and the prediction is that this will grow to 30 percent in the early 1990s. Curiously enough, it was the Japanese – particularly the perceptive Akio Morita, chairman and co-founder of Sony – who apparently first drew attention to the process.

The dangers of hollowing-out are obvious. Outsourcing shifts skills abroad and ends up boosting someone else's skills base. As C. J. van der Klugt, vice-chairman of Philips, the Dutch consumer electronics company, explains, "First you move the industrial part to the Far East. Then the development of the product goes there, because each dollar you pay to the overseas supplier is 10 cents you're giving him to develop new devices, new concepts, to compete against you."

Re-labeling can be immensely profitable to companies in the short term: Caterpillar, General Electric, RCA, Kodak, Honeywell and other

US firms have done good business importing Japanese, Korean and other Far East-made goods and sticking their own names on them. But the strategy can easily backfire when the market and distribution system reaches a critical level and the foreign suppliers decide to go it alone using their own labels.

Likewise, too many US companies are concerned about short-term profitability and the immediate interests of their shareholders to worry much about the long-term damage being done to the US economy. Long-term planning and finance is one of the strengths of the Japanese. Outsourcing or hollowing-out is clearly no real answer to America's manufacturing problem: US companies must invest in human and physical resources, raise productivity and promote innovation at home in order to ensure enduring economic success.

High technology could halt – and even reverse – the hollowing-out of US manufacturing. High-tech is reinventing the factory, as we saw in chapter 6. The technology exists for the implementation of flexible manufacturing systems (FMS) and computer-integrated manufacturing (CIM). These technologies reduce still further the labor content of manufactured goods, thus obviating the need to shift production to cheap-labor countries. They also make manufacturing more flexible: this enables companies to bring their plants back to locations nearer major markets.

But FMS and CIM require the investment of massive resources and the full commitment of US managers – a commitment which too often they are not willing to make. For instance, a recent study of 90 manufacturing companies in Pennsylvania by the Technology Management Center painted a depressing picture of outdated factories and ignorant, frightened and confused managers. Current US examples of fully implemented FMS projects number no more than a few dozen, and overseas rivals are catching up fast.

Market research firms like Dataquest and Yankee Research Group are predicting a big increase in spending on the "factory of the future" by 1990. But most of this spending is by big companies already in the factory automation game – companies like GM, Boeing, Deere, Hughes Aircraft and Allen-Bradley. Smaller job shops – in many ways the backbone of the US metal-bashing industry – rarely figure in such studies.

Yet there are some impressive examples of US companies fighting back against the Japanese. In Detroit, of course, Ford, GM and Chrysler have worked hard in recent years to approach the productivity levels obtained in Japan. IBM has been striving to source all major components and peripherals in the USA – and has succeeded in producing them at competitive cost. Tektronix in oscilloscopes, Litton Industries in microwave ovens, Black & Decker in power tools and 3M in videotape are all examples of US companies vigorously taking on and pushing back the Japanese competition. Against the odds, these companies have risked all in order to fight back.

The risks of not fighting back may be even greater. Although some academics and politicians argue that current trends in international trade are all part of an inevitable process marking America's transition to some sort of post-industrial society, it is clear that ceding certain high-volume markets to the Japanese and Far East competition pretty soon leads to the ceding of yet more markets, and of more strategically important markets.

This is what might well be happening in semiconductors and thus in high technology as a whole. US chipmakers started the rot by mostly giving up on standard memory chips or dynamic RAMs quite early on in the 1980s. Then the Japanese boosted capital spending, cut prices and moved into other chip markets such as microprocessors and VLSI chips. US semiconductor firms then began a further retreat into merely making custom chips or application-specific integrated circuits (ASICs). All this occurred over the fairly short period of less than ten years since the arrival of the first Japanese chips in 1978.

By 1986 Japan had the world's top three chipmakers – NEC, Hitachi and Fujitsu. It had also overtook the USA in semiconductor production, and was about to replace the USA as leader in the worldwide chip market worth around $25 billion. Back in 1981 the USA had had what seemed to be an unassailable lead, with over 60 percent of the world market and nearly 60 percent of world chip production; but only five years later analysts were predicting outright Japanese domination by the year 2000 and were seriously asking whether it was now too late to save the US semiconductor industry – despite the famous and much-debated US–Japan accord signed in July 1986. There was even talk of direct federal aid for the ailing chipmakers, in the form of the "Sematech" project, proposed by the SIA in October 1986.

Semiconductors are the bedrock of the entire electronics and computing industry. National success in chips determines national success in a whole range of strategic, high-tech industries such as defence, aerospace and telecommunications. Without a leading-edge semiconductor industry, US manufacturers of cars and consumer durables will be at a disadvantage – perhaps even doomed.

America is at the crossroads, and it faces a crucial choice; either it can take the road of perpetual decline like the UK, or it can make the fundamental changes needed to regain the highway of world leadership. There should be a major national debate on the issue, but at present the silence on this subject from Washington is terrifying.

What Will Japan Do Next?

The successful launch of Japan's H-1 rocket in August 1986 could not have provided a better paradigm of the current state of play between Japan, Europe and the USA in the international high-tech race. While the USA was left without a satellite launch system after the *Challenger* space shuttle disaster and Europe's *Ariane* rocket was grounded after a succession of failures, the H-1 launch went off without a hitch from Japan's space center on the island of Kyushu. Japan now looks like becoming a major player in yet another industry of the future – space.

Predicting what Japan will do next can hardly hope to be an exact science. Predicting even the medium-term future in high-tech is difficult enough, and reading the minds of the Japanese in this area is doubly difficult. And yet, unlike the West, Japan has formulated and begun various national plans and projects in recent years which give us many useful clues as to what industries the Japanese have targeted for takeover. Two in particular are featured here – the Human Frontiers Plan to boost basic research, and the national Sigma Project to produce world-class software.

The Japanese in the postwar period have developed a genius for production engineering and the adaptation of other people's technology. Their strength has been in taking ideas from elsewhere and developing them into commercial products which are then produced more efficiently and are of higher quality. Japan's expertise has been in applications and in applying existing technology, mostly first developed in the West. In the past, this led to allegations that the Japanese were simply "copycats." More recently, there have been suggestions that the Japanese are good "imitators" but lack the originality to become good "inventors."

Now Japan faces problems on three fronts. First, the country is running out of technology to import. Most of the key innovations in, for example, consumer electronics have already been borrowed from the West and there seems little else in the pipeline. Second, Japan faces growing competition in terms of labor costs and manufacturing expertise from the so-called "Four Tigers" of the East – South Korea, Taiwan, Hong Kong and Singapore. (Behind them, ready and waiting, stand the "Gang of Five" – Malaysia, Indonesia, India, Thailand and the Philippines.) Third, the recent dramatic rise in the value of the yen has put renewed pressure on the Japanese to develop more of their own ideas in order to stay ahead and remain competitive.

The Human Frontiers Plan, announced at the Tokyo economic summit in May 1986, aims to boost the funding of basic research in artificial intelligence, biotechnology, new materials and alternative energy sources. There is talk of Japanese government financial support to the tune of $2.8 billion over 20 years, and it is said that Human Frontiers will look at such basic human processes as ageing, memory and metabolism. But some skeptics have claimed that the project is something of an international PR stunt. Japanese industrialists will not see much in it for them and therefore won't be willing to contribute.

Japanese companies are now spending more anyway on basic R&D, taking over that role from the somewhat hidebound and traditional Japanese universities. Market researchers Dataquest, in a report published in February 1986, said that 76 basic research labs will have been opened by Japanese electronics companies alone in the period 1984–8. Since the Japanese can no longer rely on imitating or refining technology imported from overseas, firms like NEC, Hitachi and Canon are urgently seeking to establish themselves in leading-edge R&D. As *New Scientist* commented in a special survey of Japanese science (March 21, 1985), "Japanese industry has become the guardian of pure research."

According to a government paper published in December 1985 by its Science and Technology Agency, Japanese companies are now spending

77 percent of the national R&D budget – which was itself up from $24 billion in 1983 to $30 billion in 1985. Yet that national total is surprisingly low by international standards – the Japanese reputation for being imitators rather than originators still has some basis in fact.

Japan has a long-running "creativity problem" marked by a shortage of scientists with original ideas as opposed to engineers with production skills. Some have said this is because the notion of individual originality is at odds with the hierarchical discipline and collective loyalty of Confucian philosophy – it seems that Confucius cannot be reconciled with computer programming! That may or may not be true, but Japanese researchers do publish far fewer scientific papers than their US and European counterparts, and Japan boasts few Nobel Prize winners. The *quality* of much Japanese research has traditionally not been so good, but it is improving. The December 1985 government paper suggested various ways of beefing-up Japanese R&D, and Japanese companies have recently begun to recruit top Asian brains, for example, to its research labs.

Computer software has long been a major Japanese weakness, but now, through a new national program and numerous company programs, the Japanese are making a determined attempt to overcome this. The Sigma Project, started in October 1985, is worth $250–$300 million and runs for five years. Based in Tokyo's Akihabara electronics district, Sigma aims to achieve national software compatibility using Sigma OS, an advanced version of Unix. It plans to link no less than 10,000 software development workstations to the national HQ by 1990 and to build up a library of software for sale or loan.

Japanese companies have their own projects to automate programming by re-using parts of previously successful programs. Automating system software production is more difficult (if not impossible!), but system software tools are being developed by some companies. Software productivity and quality is now high in Japanese companies: NEC is probably most advanced in the development of program generators and other types of software engineering, and Toshiba has adopted a factory-like approach with its Software Workbench facility. There is a big push on in all leading Japanese companies to boost software productivity and develop bug-free software.

Mention has been made earlier of the emerging Japanese supremacy in chips and of the government-organized efforts being made to establish for Japan a world lead in all types of computer hardware. Chief of these are the National Super-Speed Project to develop high-speed computers and the Fifth-Generation Project to develop the hardware and software tools that should lead to artificial intelligence (AI) applications.

Already, the Japanese are hard on the heels of IBM: soon after the latter announced its 3090/Sierra series, NEC and Hitachi announced superior systems. Japanese computer companies are currently supplying US firms like UNISYS, Honeywell, Amdahl (49 percent owned by Fujitsu), National Semiconductor and Britain's ICL with chips, chip sets, micros, minis and superminis. Many US firms are entering into alliances and/or collaborative deals with the Japanese in which the Japanese look set to come out on top, because of their superior hardware.

In the research labs at Tsukuba Science City, Japanese firms such as NEC, Fujitsu and Hitachi are working on parallel architectures, new types of high-speed devices such as Josephson junctions, high-electron mobility transistors (HEMTs) and gallium arsenide chips (GaAs). Basing their work on the PROLOG programming language rather than LISP, favored by most Europeans, the Japanese are working on a new type of "dataflow" computer which uses parallel processing architectures.

In the electronics industry itself, the automation of printed circuit board production is well advanced and data communications networks are now revolutionizing the production of peripherals like printers. For instance, Toshiba's Ome works boasts a four-layer network in which a mainframe stores product and production plans; a supermini controls the production process; minis hold the sets of instructions for different models; and micros tell the production-line robots and assembly machines what to do. Fiber optics or optical networks are being widely used by the Japanese for communications links – especially for the transmission of data, text and graphics both to overseas plants and to facilities within Japan.

The Japanese are also determined to stay ahead in the industries where they have already triumphed, such as motorcycles, cars, cameras, consumer electronics and computer peripherals. And they plan to take the lead in the emerging areas of biotechnology and materials such as composites and the new ceramics.

In all areas of manufacturing, the Japanese plan to further simplify production processes and to make them more flexible, with the eventual goal of manufacturing in "batches of one." The increased use of "just-in-time" (JIT) inventory control, improvements in robotics and the further development of numerically controlled (NC) machine tools suggests that the Japanese will never be satisfied with existing manufacturing techniques, but will always find more to do.

In NC machine tools, for instance – where the Japanese now have a world lead – they plan the widespread use of laser metal-cutting machines, flexible machining centers and high-precision molding machines. Optical communications networks will convey the new compatible software to sophisticated automated guided vehicles (AGVs) and a new generation of autonomous robots. Japanese cleverness, resourcefulness and attention to detail show no signs of letting up.

In vehicle production – Japan produced 12.3 million and the USA only 11.7 million vehicles in 1985 – the Japanese will forge ahead with "mixed model" assembly lines. These will produce a staggering range of models and specifications for numerous, vastly different export markets. Automation will continue apace: at Nissan's Zama plant, for instance, robots are now being used not only for welding and paint-spraying, but also for fitting spare tires, batteries, windows, seats, rear doors and rear lamps – and they even fill the radiator and screenwasher, measure the wheel alignment and adjust the headlight aim! Even so, Nissan estimates that only 8 percent of operations on the final assembly line are in fact automated: there is still a long way to go.

In motorcycles – scene of one of the earliest Japanese successes – the Japanese intend to maintain world domination with constant improvements in engine performance: their motorbike engines will feature more and more power for less and less weight. Endless innovation has kept Japan way ahead in this field, but a steady fall in the total world market has seen the resourceful Japanese swiftly and successfully diversify into lawn mowers, outboard motors, generators, snowmobiles, golf carts and other small-motor products. The development of overhead valve engines for improved performance and automatic decompression systems for easier starting should give the Japanese the technological edge.

The major problem facing the Japanese in the coming decade will be not to produce the goods that the world wants to buy, but to maintain access to markets in order to sell them. Japan's remarkable and sudden success has led to trade tension and increased pressure for protectionism in other countries. There have been frequent pleas to the Japanese to import more and to build more plants in overseas markets. Calls for action on Capitol Hill reached something of a crescendo in early 1985, when it was revealed that the US–Japan trade deficit had soared to $37 billion in 1984. Japan's prime minister Yasuhiro Nakasone even went on national TV to urge the Japanese people to buy more foreign goods, amid much talk in Washington of a coming "trade war" with Japan.

There was justified anger in Silicon Valley at the predatory pricing policies of Japanese chipmakers and widespread concern among US exporters with the well documented bureaucratic barriers to imports – some subtle, some not-so-subtle – created by the Japanese authorities. Congress seemed on the verge of introducing major protectionist measures, but without the President's support little came of this. Even so, in Europe there have been calls by EEC spokesmen and from companies like Philips for the Japanese to scrap their "unfair" trading practice of targeting industries for takeover by ruthless price-cutting which amounts to dumping.

With the rise of the yen, 1986 brought news from Japan of factory closures, mergers and restructuring, plant openings overseas and even outsourcing deals as Japanese companies began a fairly dramatic process of adjustment to the changing international trade regime. There were also signs of Japan opening up its internal market to a greater degree: US telecoms companies won more than $1 billion worth of orders for telecoms equipment in the summer months of 1986. Even so, Japan was still running a massive trade surplus and was maintaining its boldly expansionist stance in world markets.

Japan is a stable, prosperous, orderly and peace-minded society. The Japanese economic miracle has been a major theme – if not *the* major theme – of the postwar era. Forty years ago, most of Japan lay in ruins: now it has overtaken the USSR in terms of GDP and it looks like becoming the world's leading economic power by the end of the century.

On the other side of the coin, Japan suffers from overcrowded cities, poor infrastructure, a high rate of marriage breakups and stress-related diseases. The workaholic Japanese – allegedly unable to relax or play – are prone to alcoholism, breakdowns and suicide. The human cost of economic success is high, and there are signs that some Japanese may no longer be prepared to pay for it: the idea of lifelong loyalty and dedication to one company, for example, may be in decline.

But for the foreseeable future, it looks like nothing can stop them. Veteran Japan-watcher, Professor Ronald Dore of London's Imperial College, has little doubt that Japan will stay ahead by sustaining its supremacy in high-tech. In Japan today, Dore says, there is an emerging "optimal conjunction" of a good supply of scientists and technologists in new, well-equipped labs – people who are also currently motivated enough to stay at work until 11 pm!

Once Britain had this desire to succeed, and it became known as the "workshop of the world." Then technological leadership shifted to the USA and Germany. Now, the one to beat is Japan.

Can Europe Catch Up?

With the USA and Japan slogging it out for world high-tech supremacy, poor old Europe is caught in the crossfire and looks like being the loser. From 1945 until 1978, Europe ran a positive trade balance in electronics, computers and telecoms equipment. But with its vitality during the postwar decades on the wane, in 1979 Europe slipped into deficit in high-tech trade, and it has remained there ever since. As further opportunities were lost, the Continent fell to a poor third place in the world high-tech stakes.

Now time is running out for Europe. With the rise of the Far East and Third World nations that seem better able to compete, people are asking whether Europe can catch up and whether there is anything it can do to prevent itself from sinking to third-, fourth- or even fifth-rate economic status. In short, can Europe *survive* at all as a major force in high-tech?

Europe's trading position is dire: a $1.7 billion surplus on high-tech goods in 1975 was converted into a $2 billion deficit by 1982. By 1984 the deficit had grown to $5 billion, and by 1986 it reached $10 billion. Between 1970 and 1982, the EEC's share of the world market in high-tech goods fell to 17 percent, while the US share grew to 36 percent and the Japanese to 38 percent. Meanwhile, total industrial production grew by 16 percent in the USA and 26 percent in Japan over the period 1973–83, whereas in Europe it grew by only 8 percent – although continental productivity growth did outstrip that of the USA.

In computers, the USA and Japan sell Europe four times as much computer equipment as they buy from it. European IT companies have only 40 percent of their own market, and IBM outsells all European computer companies put together – on their own turf. The UK's ICL, France's Bull and West Germany's Siemens each has less than 2 percent of the world market, although Norsk Data of Norway and Nixdorf of West Germany are growing success stories. Italy's Olivetti is also doing well with its aggressive international strategy.

The history of European computing is a long litany of lost opportunities. Apart from the UK having the first mainframe, UK companies once had world leadership in robotics and NC machine tools – leadership that has long since been surrendered to the Japanese. Some of the first specialist applications of CAD were developed in Europe (particularly at Cambridge, England), but now, it is the USA that has a lead in CAD systems. European computer companies for too long lived an easy life off regular work for national PTTs, lucrative government contracts and state R&D grants. This feather-bedded the large established firms, stifled small entrepreneurial companies and did Europe no good in the long run.

Now, Europe has no supercomputer to offer, a few struggling mainframe makers, a small presence in minicomputers and a handful of microcomputer makers – only Olivetti in this section is world class. But Olivetti is teamed with the USA's AT&T and buys in from the USA, Japan and Hong Kong. Britain's ICL uses Fujitsu technology, France's

Bull uses NEC and Honeywell kit, while Siemens also relies heavily on Fujitsu. As Alan Cane of London's *Financial Times* points out, "Europe seems to have played almost no part in the development of the major current trends in data processing." Where was the European Compaq, Apple, Tandem, Stratus or Convex, he asks?

Software was supposed to be a European strongpoint, but the Continent is actually losing market share; it has been estimated that 70 percent of the software in use in Europe is of US origin. European companies missed out on the packaged software revolution, and too many European software specialists have been tied up developing telecoms systems for national PTTs. Europe should benefit from the trend to custom software, but many analysts are skeptical. Standardizing on OSI in Europe in the early 1990s will help Europe's competitors as much as European companies.

In chips, the European share of the world market fell to less than 10 percent by the mid-1980s, although in late 1986 there were signs of a modest pickup to around 12–13 percent. Only one European company – Philips – is in the world Top Ten of chip makers; but it is placed at No. 8, and no less than two-thirds of its sales are generated by its US subsidiary, Signetics. The European market for chips constitutes only about 18 percent of the world market – down from 30 percent in the 1970s – and European chipmakers have only about 3 percent of the world market outside Europe.

European chipmakers have two main handicaps: high production costs and sluggish home markets. But they have been slow to respond to new trends. The UK firm of Ferranti provides a classic example of how to fritter away a five-year lead in gate arrays or semi-custom chips. Once a world leader with 30 percent of the market, Ferranti has now fallen behind Fujitsu, Motorola, LSI Logic and NEC. Yet European manufacturers did weather the 1985 chip slump better than most, precisely because of their concentration on semi-custom chips and discrete components.

Europe is in danger of falling further behind in telecommunications. European telecoms companies – mostly state-owned PTTs – still had 25 percent of the world market in 1985. But their historical advantages of protected home markets and "captive" sales to Third World nations are dwindling fast. The Balkanization of Europe's telecoms markets has led to the massive duplication of R&D costs – Europe has no less than ten rival digital public exchange switching systems, each of which has cost between $500 million and $1 billion to develop.

In a bid to protect their own "national champions," European governments have created a chaos of varying technical standards, rules, regulations and charges. As a result, Europe has a phone system that is expensive, unreliable and a cause of much frustration to business users. Despite new measures of liberalization and privatization (especially in the UK, but also now in France), European PTTs have been slow to react to technological change and slow to develop new on-line information services and VANs.

While individual European companies like Sweden's L. M. Ericsson and West Germany's Siemens have done well selling abroad, especially in the USA, most European telecoms companies have failed abysmally in world markets. Alcatel, the tie-up between the state-owned CGE of France and ITT of the USA, launched early in 1987, is a step in the right direction of ending fragmentation and may lead to a much-needed restructuring of the European telecoms industry, particularly if the mooted links with Spain's Telefonica and Belgium's Société Generale go ahead.

Europe has some very successful companies in sectors such as cars, pharmaceuticals, chemicals and food processing. Firms like Daimler–Benz, Volvo, BMW, Bosch, Bobst, Pomogalski, Electrolux, Glaxo, Hoechst, Bayer, Ciba-Geigy, ICI, Fisons, Tetra-Pak and Heineken are world class. There have been some notable "turnarounds" – of Fiat, Olivetti, Peugeot, Jaguar and Bull. There have also been two very successful examples of European cooperation: the European Airbus and the *Ariane* space rocket (until recent failures). But Europe has also experienced too many failures and wasted too many opportunities.

Despite the existence of the EEC, Europe is still very much an Uncommon Market. A series of fragmented markets means that European companies cannot achieve the economies of scale needed to compete effectively in world markets – although some, such as ASEA and Ericsson of Sweden and Bobst of Switzerland, have actually succeeded from a small home base. Visible and not-so-visible trade barriers are created by a patchwork of rules and regulations on safety and design which not only lead to border delays but also add 8–12 percent to the costs of goods traded *within* the Community, according to some economists.

The policy of "national champions" makes cross-border cooperation difficult – indeed, European nations and companies have consistently failed to cooperate successfully. In the 1960s Euratom was supposed to coordinate European efforts in atomic energy, but it flopped. In the early 1970s Philips of Holland, Siemens of West Germany and CII/Bull of France joined together to form Unidata, which was supposed to become the united European rival to IBM. But it fell apart when the French defected to Honeywell instead. There was also as series of disastrous trans-European company mergers in the 1960s and 1970s which did not last (such as Agfa–Gaevert, Dunlop–Pirelli, VFW–Fokker and Hoechst–Hoogovens). Europeans keep *talking* about the need for European alliances and so on, but in practice they tend to go into partnership with Japanese and US companies rather than with each other. In this way, they hope to gain access to the vast US market or to obtain Japanese know-how.

But there is a deeper problem in Europe: "Eurosclerosis." Professor Herbert Giersch, head of the Kiel Institute of World Economics, coined the term to describe the hardening of the Continent's economic arteries owing to rigid labor markets, state bureaucratic interference in business, conservative bankers and the lack of venture capital, high taxes, poor management and outdated education systems, to name but a few factors. A measure of the rigidity of European business is provided by the economist Henry Ergas of the Paris-based OECD: he points out that Europe's largest high-tech companies today are the same as they were 35 years ago! In the USA, turnover at the top has been much higher, and in Japan the situation has almost changed completely.

One aspect of Eurosclerosis is Europe's perennial problem of not being able to turn the fruits of R&D into commercial products. Even though an amazing 66 percent of the EEC budget goes on agricultural subsidies and only 3 percent on technology, R&D spending in Europe as a proportion of GNP is roughly comparable to that of the USA and Japan. But too high a proportion is spent in old-fashioned government labs with a poor record of innovation, and too little effort is put into making the leap from lab to marketplace by targeting R&D efforts for commercial applications.

Another is the shortage of entrepreneurs. Despite the two world wars, Europe still has plentiful supplies of human, technical and financial capital. But it lacks the entrepreneurs, the managers and the people with the marketing skills needed to build world-class companies and to go out boldly and tackle competitive world markets. "The number of potential growth companies in Europe run by people with a growth mentality is very small," says Raymond Appleyard, head of the EEC's innovation division.

As we saw in chapter 1, official EEC efforts to revitalize European high-tech have concentrated on three research programs: ESPRIT, RACE and EUREKA. The key point about the $1.3 billion ESPRIT program, begun in 1982, is that it is aimed at strengthening pre-competitive basic research and is not really geared to producing commercial products. However, it is helping build links between companies and academic research institutions. RACE, also started in

1982, is an EEC-inspired effort to boost cooperation in telecoms. Four companies – CIT Alcatel, Italtel, Siemens and Plessey – were originally involved, but recently even IBM has expressed an interest, pointing perhaps to joint efforts in the 1990s.

EUREKA was launched in 1985 by the French specifically as an alternative attraction to SDI/Star Wars, but it soon broadened out into a wide-ranging program of IT research. Unlike ESPRIT, EUREKA is supposed to be more of a "downstream" operation focusing on commercial applications. EUREKA has attracted plenty of applicants for funding, raising suspicions that it might be subsidizing projects that would have been undertaken anyway. Many of them read like a list of European lost opportunities, and some seem too abstruse to make an early leap out of the lab.

The point has been frequently made – for example, in the recent book edited by Margaret Sharp, *Europe and the New Technologies* (Frances Pinter, London, 1985) – that direct government intervention in the form of schemes like ESPRIT and EUREKA is unlikely on its own to solve Europe's high-tech problem. The real answer may well be to abandon the old policies of protected "national champions" completely and to create a vigorous, competitive economic environment which might result in the emergence of a new generation of entrepreneurs determined to achieve world market leadership. But that could take time – and time is not on Europe's side.

Two promising private sector microchip initiatives are the Philips–Siemens Megaproject and the pan-European start-up, European Silicon Structures, or ES2. Megaproject is a $500 million three-year joint effort designed to leapfrog the Japanese and develop the next generation of memory chips. It was going for 1 megabit chips, but since the Japanese have already announced their versions, Megaproject is now having to go for 4 megabit chips.

ES2 is an Anglo-French venture with HQ in Munich, West Germany. Founders Jean-Luc Grand-Clement (ex-Motorola and National Semiconductor), Robb Wilmot (ex-Texas Instruments) and Robert Heikes (ex-National Semiconductor) raised $60 million from seven European companies – Olivetti, British Aerospace, Philips, Brown Boveri, Telefonica, Bull and Saab-Scania – with the express purpose of putting Europe back on the leading edge of chip technology. ES2 will go for sophisticated custom chips, produced using silicon compilers and electron beam lithography.

In fact, there are even further signs of a European chip renaissance in the shape of other start-ups, including Belgium's Mietec and Scotland's Integrated Power Semiconductors. France's Thomson under Jacques

Noels (ex-Texas Instruments) is spending heavily – by buying the assets of Mostek in 1985, for instance. Italy's chipmaker, SGS-Ates, run by ex-Motorola man Pasquale Pistorio, is building major new chip plants in Milan, Singapore and Phoenix (Arizona).

Behind these investments in semiconductor production facilities lies the realization by perceptive Europeans that competitiveness in a whole range of products, from cars to domestic appliances, is now determined by the power and the abilities of the chips they contain. As product life-cycles shorten, early access to the very latest chip technology is a vital ingredient of commercial success.

One thing is clear: if Europe continues in the same old, bad ways of protected "national champions" in fragmented markets, it will have no future at all in high-tech. Ossified systems, which featherbed established companies or create cosy monopolies, and which prize status and security above performance and risk, are highly vulnerable to high-tech attack from outside. Continuing technological innovation and risk-taking entrepreneurship – and an economic and social climate that helps them both to flourish – are probably the best hopes for Europe's survival in the face of worldwide competition.

Selected Further Reading

Chapter 1 Introduction

Stan Augarten, *Bit by Bit: An Illustrated History of Computers and Their Inventors* (Ticknor & Fields, New York, 1984, and Allen & Unwin, London, 1985).

Jeremy Bernstein, *The Analytical Engine: Computers – Past, Present and Future* (Morrow, New York, 1981).

Joel Shurkin, *Engines of the Mind: A History of the Computer* (W. W. Norton, New York, 1984).

Tom Forester (ed.), *The Information Technology Revolution* (Basil Blackwell, Oxford, England, 1985 and MIT Press, Cambridge, Mass., 1985).

Lynn M. Salerno, *Computer Briefing: The Concise Update on the Latest Developments* (John Wiley, New York, 1986).

Barry Jones, *Sleepers, Wake! Technology and the Future of Work* (Oxford University Press, Melbourne, Australia, 1982, and Wheatsheaf, Brighton, England, 1982).

Gordon Pask and Susan Curran, *Micro Man: How Computers are Revolutionising Our Lives* (Century, London, 1982).

Alvin Toffler, *The Third Wave* (Morrow, 1980, and Bantam Books, New York, 1981; Collins, 1980, and Pan Books, London, 1981).

J. David Bolter, *Turing's Man: Western Culture in the Computer Age* (Duckworth, London, 1985, and Penguin, New York and London, 1986).

Michael L. Dertouzos and Joel Moses (eds), *The Computer Age: A Twenty-Year View* (MIT Press, Cambridge, Mass., 1979).

Simon Nora and Alain Minc, *The Computerization of Society* (MIT Press, Cambridge, Mass., 1980).

Nils Bjorn-Anderson, Michael Earl, Olav Holst and Enid Mumford (eds), *Information Society: For Richer, For Poorer* (Elsevier, Amsterdam-North Holland, 1982).

Liam Bannon, Ursula Barry and Olav Holst (ed), *Information Technology: Impact on the Way of Life* (Tycooly, Dublin, Eire, 1982).

Dale Whittington (ed.), *High Hopes for High Tech* (University of North Carolina Press, 1986).

Chapter 2 The Computer Revolution

T. R. Reid, *The Chip: How Two Americans Invented the Microchip and Launched a Revolution* (S & S Publishing, North Carolina, 1985). Published in the UK as *Microchip: The Story of a Revolution and the Men Who Made It* (Collins, London, 1985).

Ernest Braun and Stuart Macdonald, *Revolution in Miniature: The History and Impact of Semiconductor Electronics* (Cambridge University Press, Cambridge, England, and New York (2nd edn), 1982).

Tom Forester (ed.), *The Microelectronics Revolution* (Basil Blackwell, Oxford, England, 1980, and MIT Press, Cambridge, Mass., 1981).

Joseph Weizenbaum, *Computer Power and Human Reason: From Judgement to Calculation* (W. H. Freeman, San Francisco, 1976, and Penguin Books, London and New York, 1984).

Edward A. Feigenbaum and Pamela McCorduck, *The Fifth Generation: Artificial Intelligence and Japan's Computer Challenge to the World* (Addison-Wesley, Reading, Mass., 1983, and Michael Joseph, London, 1984).

Margaret A. Boden, *Artificial Intelligence and Natural Man* (Harvester, Brighton, England, 1977, and Basic Books, New York, 1981).

Patrick H. Winston and Karen A. Prendergast (eds), *The AI Business: Commercial Uses of Artificial Intelligence* (MIT Press, Cambridge, Mass., and London, England, 1984).

John Haugeland, *Artificial Intelligence: The Very Idea* (MIT Press, Cambridge, Mass., 1985).

Eugene Charniak and Drew McDermott, *Introduction to Artificial Intelligence* (Addison-Wesley, Reading, Mass., 1985).

Donald Michie and Rory Johnston, *The Creative Computer: Machine Intelligence and Human Knowledge* (Viking, New York, 1984, and Penguin Books, Harmondsworth, England, 1985).

Karamjit S. Gill (ed.), *Artificial Intelligence for Society* (John Wiley, New York and Chichester, England, 1986).

Masoud Yazdani (ed.), *Artificial Intelligence: Principles and Applications* (Chapman & Hall, London, 1986).

Hubert Dreyfus and Stuart Dreyfus, *Mind Over Machine* (Macmillan/Free Press, New York, 1986).

Chapter 3 Silicon Valley: Home of High-Tech Man

Everett M. Rogers and Judith K. Larsen, *Silicon Valley Fever: Growth of High Technology Culture* (Basic Books, New York, 1984, and George Allen & Unwin, London, 1985).

Michael Moritz, *The Little Kingdom: The Private Story of Apple Computer* (Morrow, New York, 1984).

Michael S. Malone, *The Big Score: The Billion Dollar Story of Silicon Valley* (Doubleday, New York, 1985).

John W. Wilson, *The New Venturers: Inside the High-Stakes World of Venture Capital* (Addison, Wesley, Reading, Mass., 1985).

Regis McKenna, *The Regis Touch* (Addison-Wesley, Reading, Mass., 1985).

Stephen T. McClennan, *The Coming Computer Industry Shakeout: Winners, Losers and Survivors* (John Wiley, New York, 1984, and Chichester, England, 1985).

Tracy Kidder, *The Soul of a New Machine* (Little, Brown, Boston, 1981, and Penguin, Hardmondsworth, England, 1982).

Frank Rose, *Into the Heart of the Mind: A Quest for Artificial Intelligence* (Harper & Row, New York, 1984, and Century, London, 1985).

Jay Tuck, *High-Tech Espionage: How the KGB Smuggles Nato's Strategic Secrets to Moscow* (Sidgwick & Jackson, London, 1986).

Chapter 4 The Telecommunications Explosion

Susan J. Tolchin and Martin Tolchin, *Dismantling America: The Rush to Deregulate* (Houghton Mifflin, New York, 1984).

Harry M. Shooshan III (ed.), *Disconnecting Bell: The Impact of the AT&T Divestiture* (Pergamon, Elmsford, NT, 1984).

W. Brooke Tunstall, *Disconnecting Parties: Managing the Bell System Break-Up – An Inside View* (McGraw-Hill, New York, 1985).

Jeremy Tunstall, *Communications Deregulation: The Unleashing of America's Communications Industry* (Basil Blackwell, Oxford, England, and New York, 1986).

Larry Kahaner, *On the Line* (Warner Books, New York, 1986).

Richard M. Neustadt, *The Birth of Electronic Publishing: Legal and Economic Issues in Telephone, Cable and Over-the-Air Teletext and Videotext* (Knowledge Industry Publications, White Plains, NY, 1982).

Brian Wenham (ed.), *The Third Age of Broadcasting* (Faber, London, 1982).

Timothy Hollis, *Beyond Broadcasting: Into the Cable Age* (British Film Institute, London, 1984).

Christopher M. Byron, *The Fanciest Dive* (W. W. Norton, New York, 1986).

James Martin, *Viewdata and the Information Society* (Prentice-Hall, Englewood Cliffs, NJ, 1982).

Michael Aldrich, *Videotex: Key to the Wired City* (Quiller Press, London, 1982).

Chapter 5 Personal Computing

Paul Freiberger and Michael Swaine, *Fire in the Valley: The Making of the Personal Computer* (Osborne/McGraw-Hill, Berkeley, Cal., 1984).

Neil Frude, *The Intimate Machine: Close Encounters With the New Computers* (Century, London, 1983, and New American Library, New York, 1983).

Sherry Turkle, *The Second Self: Computers and the Human Spirit* (Simon & Schuster, New York, 1984, and Granada, London, 1984).

Jack M. Nilles, *Exploring the World of the Personal Computer* (Prentice-Hall, Englewood Cliffs, NJ, 1982).

Robin Bradbeer, *The Personal Computer Book* (2nd edn) (Gower, Aldershot, England, 1982).

Alfred Glossbrenner, *The Complete Handbook of Personal Computer Communications* (St Martins Press, New York, 1983).

Ben Schneider Jr, *My Personal Computer and Other Family Crises* (Macmillan, New York, 1985).

Raymond S. Nickerson, *Using Computers: The Human Factors of Information Systems* (Bradford Books/MIT Press, Cambridge, Mass., 1986).

Christopher Dunkley, *Television Today and Tomorrow: Wall-to-Wall Dallas?* (Penguin, Hardmonsworth, England, and Viking, New York, 1985).

Seymour Papert, *Mindstorms: Children, Computers and Powerful Ideas* (Basic Books, New York, 1980).

Tim O'Shea and John Self, *Learning and Teaching with Computers* (Harvester, Brighton, England, 1983).

Margaret Graham, *RCA and the Videodisc: the Business of Research* (Cambridge University Press, Cambridge, England, 1986).

Chapter 6 Factories of the Future

Norbert Wiener, *The Human Use of Human Beings* (Houghton Mifflin, New York, 1950).

John Diebold, *Automation: The Advent of the Automatic Factory* (Van Nostrand, New York, 1952).

Robert U. Ayres and Steven M. Miller, *Robotics: Applications and Social Implications* (Ballinger, Cambridge, Mass., 1983).

Igor Aleksander and Piers Burnett, *Reinventing Man: The Robot Becomes Reality* (Kogan Page, London, 1983, and Penguin, Harmondsworth, England, 1984; Holt, Rinehart & Winston, New York, 1984).

Peter B. Scott, *The Robotics Revolution* (Basil Blackwell, Oxford, England, and New York, 1984).

D. J. Todd, *Fundamentals of Robot Technology* (Kogan Page, London, 1986).

Richard J. Schonberger, *Japanese Manufacturing Techniques* (Free Press, New York, 1982).

Robert U. Ayres, *The Next Industrial Revolution: Reviving Industry Through Innovation* (Ballinger, Cambridge, Mass., 1984).

Alan Altshuler, Martin Anderson, Daniel Jones, Daniel Roos and James Womack, *The Future of the Automobile* (MIT Press, Cambridge, Mass., 1984).

Chapter 7 The Electronic Office

Susan Curran and Horace Mitchell, *Office Automation* (Macmillan, London, 1982).
Denis Jarrett, *The Electronic Office* (Gower, Aldershot, England, 1982).
Andrew Doswell, *Office Automation* (Wiley, New York, 1983).
John J. Stallard, E. Ray Smith and Donald Reese, *The Electronic Office: A Guide for Managers* (Dow-Jones Irwin, New York, 1983).
John Steffens, *The Electronic Office: Progress and Problems* (Policy Studies Institute, London, 1983).
Paul A. Strassman, *Information Payoff: The Transformation of Work in the Electronic Age* (Free Press, New York, 1985).
Steven Lambert (ed.), *CD ROM: The New Papyrus* (Microsoft Press, Redmond, Washington, 1986).
R. A. Hirschheim, *Office Automation: A Social and Organizational Perspective* (John Wiley, Chichester, England and New York, 1985).

Chapter 8 Micros and Money

Effects of Information Technology on Financial Services, Office of Technology Assessment (OTA) report (US Congress, Washington, DC, 1984).
John Marti and Anthony Zeilinger, *Micros and Money* (Policy Studies Institute, London, 1982).
Amin Rajan, *New Technology and Employment in Insurance, Banking and Building Societies* (Gower, Aldershot, England, 1984).
Geoffrey Cooke, *Technology and Employment in the London Clearing Banks* (Banking Information Service, London, 1986).
John Plender and Paul Wallace, *The Square Mile: A Guide to the New City of London* (Century, London, 1985).
Adrian Hamilton, *The Financial Revolution* (Viking, London, 1986).

Chapter 9 Key Problems for High-Tech Society

James Robertson, *Future Work: Jobs, Self-Employment and Leisure After the Industrial Age* (Gower, Aldershot, England, 1985).
Charles Handy, *The Future of Work* (Basil Blackwell, Oxford and New York, 1984).
Gail Garfield Schwartz and William Neikirk, *The Work Revolution* (Rawson, New York, 1983).
Stephen G. Peitchinis, *Computer Technology and Employment* (St Martins Press, New York, 1983).
Harley Shaiken, *Work Transformed: Automation and Labor in the Computer Age* (Holt, Rinehart and Winston, New York, 1985).
Colin Gill, *Work, Unemployment and the New Technology* (Polity Press, Oxford, England, 1985).

Donald Leach and Howard Wagstaff, *Future Employment and Technological Change* (Kogan Page, London, 1986).
P. K. Marstrand (ed.), *New Technology and the Future of Work and Skills* (Frances Pinter, London, and Dover, NH, 1984).
Barry Wilkinson, *The Shopfloor Politics of New Technology* (Heinemann, London, 1983).
David A. Buchanan and David Boddy, *Organisations in the Computer Age* (Gower, Aldershot, England, 1983).
Colin Lewis, *Managing With Micros*, Third Edition (The Economist/Basil Blackwell, Oxford, England, 1986 and New York, 1987).
Ed Rhodes and David Wield (eds), *Implementing New Technologies* (Open University/Basil Blackwell, Oxford, England, 1985).
Eric Batstone and Stephen Gourlay, *Unions, Unemployment and Innovation* (Basil Blackwell, Oxford, England, 1986).
Steven Levy, *Hackers: Heroes of the Computer Revolution* (Doubleday, New York, 1985).
Hugo Cornwall, *The New Hacker's Handbook* (Century, London, 1986).
Bill Landreth, *Out of the Inner Circle* (Microsoft Press, Bellvue, Washington, 1985).
David Burnham, *The Rise of the Computer State* (Random House, New York and Weidenfeld & Nicolson, London, 1983).
Duncan Campbell and Steve Connor, *On the Record: Surveillance, Computers and Privacy* (Michael Joseph, London, 1986).
Tony Solomonides and Les Levidow (eds), *Compulsive Technology: Computers As Culture* (Free Association Books, London, 1985).
Theodore Roszak, *The Cult of Information* (Pantheon Books, New York, 1986).
Frank Webster and Kevin Robins, *Information Technology: A Luddite Analysis* (Ablex, Norwood, New Jersey, 1986).

Chapter 10 Conclusion

William J. Abernathy, Kim B. Clark and Alan M. Kantrow, *Industrial Renaissance: Producing a Competitive Future for America* (Basic Books, New York, 1983).
Robert B. Reich, *The Next American Frontier* (Times Books, New York, 1983, and Penguin, Harmondsworth, England, 1984).
Michael J. Piore and Charles F. Sabel, *The Second Industrial Divide: Possibilities for Prosperity* (Basic Books, New York, 1984).
Richard J. Schonberger, *World Class Manufacturing* (Free Press, New York, 1986).
Herman Kahn and Thomas Pepper, *The Japanese Challenge* (Harper & Row, New York, 1979).
Ezra F. Vogel, *Japan as Number One* (Harper & Row, New York, 1980).

William Ouchi, *Theory Z: How American Business Can Meet the Japanese Challenge* (Addision-Wesley, Reading, Mass., 1981, and Avon Books, New York, 1982).

Richard Tanner Pascale and Anthony G. Athos, *The Art of Japanese Management* (Simon & Schuster, New York, 1981 and Penguin, Harmondsworth, England, 1982).

Lester C. Thurow (ed.), *The Management Challenge: Japanese Views* (MIT Press, Cambridge, Mass., 1985).

James C. Abegglen and George Stalk Jr, *Kaisha: The Japanese Corporation* (Basic Books, New York, 1986).

David Halberstam, *The Reckoning* (Morrow, New York, 1986).

Margaret Sharp (ed.), *Europe and the New Technologies* (Frances Pinter, London, 1985).

Ian Mackintosh, *Sunrise Europe: The Dynamics of Information Technology* (Basil Blackwell, Oxford, England, 1986).

Index

Bold type signifies a defining entry, italics an explanatory section.